SPATIAL DATA INFRASTRUCTURES IN CONTEXT

NORTH AND SOUTH

SPATIAL DATA INFRASTRUCTURES IN CONTEXT

NORTH AND SOUTH

EDITED BY
ZORICA NEDOVIĆ-BUDIĆ
JOEP CROMPVOETS
YOLA GEORGIADOU

CRC Press
Taylor & Francis Group
Boca Raton London New York

CRC Press is an imprint of the
Taylor & Francis Group, an **informa** business

CRC Press
Taylor & Francis Group
6000 Broken Sound Parkway NW, Suite 300
Boca Raton, FL 33487-2742

Printed in the United States of America on acid-free paper
10 9 8 7 6 5 4 3 2

International Standard Book Number: 978-1-4398-2802-1 (Hardback)

Visit the Taylor & Francis Web site at
http://www.taylorandfrancis.com

and the CRC Press Web site at
http://www.crcpress.com

Contents

Foreword

The ambition to develop accessible and comprehensive information systems for territorial matters is widespread in the governmental sector, ranging from large supranational bodies (e.g., the European Union) to municipalities or local communities across the globe. In the course of the past two decades, considerable effort and resources have been devoted to the introduction of spatial data processing capabilities and to the construction of geographic information systems (GISs) in a large number of governmental bodies. The important level of uptake of GIS technology across different levels of government is an indicator of the large expectations of the technology, and information systems have been created to facilitate management and use of geographic information. In the wake of the so-called information technology revolution, policy makers, administrators, and various stakeholders (among them not in the least citizens) have indeed grown accustomed to the promise and usability of exponentially enhanced technological capacity for data and information storage, processing, and exchange.

Most current GISs are designed to serve specific organizations or projects. As a result of the limited and isolated scope of development, the accessibility and interoperability of the information systems are not optimal. Barriers encountered can be technological—related to the characteristics of spatial data (geometry, semantics)—or nontechnological. The latter comprises legal (owner rights, liability, copyrights, compatibility with European Union public sector information directive), organizational, financial, and economical aspects. In the future, these nontechnological issues will likely form the main barriers.

In order to improve accessibility, interoperability, and affordability of spatial data and information, the focus of the GI community is now increasingly shifting to the challenges associated with integrating these individual systems into a space- and time-independent continuum to support (1) public authorities and administrations at various levels, (2) thematic user communities, (3) enterprises, and (4) citizen-oriented society as a whole. A "spatial data infrastructure" (SDI) is the envisioned outcome of such endeavors. An SDI addresses both technological and nontechnological issues, ranging from the creation and maintenance of GI for a wide range of themes, technical standards and protocols, and organizational issues to data policy issues, including data access policy.

In the information society, information infrastructures are becoming the backbone of the public sector. Public administration and public policy will not be based on hierarchy, but rather on databases and information networks. In this way, the development of an SDI is expected to lead to profound public sector innovation. Classic hierarchical administrative structures will make way for networks of information. These networks of information will process and exchange information on citizens, organizations, and geographically related elements. The increasing importance of networks of information will change the identity and role of the public sector, its relations with other actors in society, and its internal processes.

In addition, the relation with databases is shifting from public organizations with databases to databases with public organizations. In the information society, the public sector will have to play new roles (e.g., collection of information in authentic sources of information). The public sector will develop new relations: Instead of classic hierarchical relations, it will operate in horizontal networks of partnership and collaboration. Tasks will be driven from user perspectives and reallocated between the public, not-for-profit, and private sectors. These new identities, roles, and relations will affect the internal processes of the public sector and its interfaces with other actors in society. Classic bureaucratic processes will have to be innovated and redesigned in order to be effective and accountable.

This book is a welcome and timely contribution to the theory and practice of SDIs, and in many respects breaks new ground in improving our understanding of the increasing relevance and value of SDIs. It also explores theoretical issues and provides empirical studies related to SDIs. I am particularly pleased that the Katholieke Universitieit Leuven has been able to play an important role in supporting this research.

Prof. Dr. Geert Bouckaert
Katholieke Universiteit Leuven
President of the European Group for Public Administration (EGPA)

Acknowledgments

This book is the result of an initiative supported by the Association of Geographic Information Laboratories in Europe (AGILE, http://www.agile-online.org). We would like to acknowledge AGILE for hosting the first workshop, "Multi- and Interdisciplinary Research on Spatial Data Infrastructure Development," in Girona, Spain, in May 2008 and Delft University of Technology for hosting the second workshop, "Theory-Based SDI Research: North and South," in Delft, The Netherlands, in June 2009. We also recognize the excellent support and commitment to the book lent by the SPATIALIST project members (www.spatialist.be).

We would like to express our great appreciation for the excellent and constructive reviews offered by our esteemed colleagues. Their thoughts and suggestions helped ensure the high quality of the individual chapters and the book as a whole. We would also like to thank Professors Geert Bouckaert and Harlan Onsrud for bracing our book with their testimonials on its relevance and contribution.

Zorica Nedović-Budić, Joep Crompvoets, and Yola Georgiadou

Foreword:

Geert Bouckaert, Katholieke Universiteit Leuven, Belgium

Afterword:

Harlan Onsrud, University of Maine at Orono

Reviews:

Margunn Aanestad, University of Oslo, Norway

Victor Bekkers, Erasmus University, Rotterdam, The Netherlands

Naomi Bloch, University of Illinois

Chip Bruce, University of Illinois at Urbana-Champaign

Nicholas Chrisman, Laval University, Quebec, Canada

Walter de Vries, University of Twente/ITC, The Netherlands

Hamid Ekbia, Indiana University, Bloomington

Sarah Elwood, University of Washington, Seattle

Jon Gant, University of Illinois at Urbana-Champaign

Raoni Guerra Lucas Rajão, University of Lancaster, United Kingdom

Niall Hayes, University of Lancaster, United Kingdom
Paul Hendriks, Radboud University Nijmegen, The Netherlands
Vincent Homburg, Erasmus University, Rotterdam, The Netherlands
Alenka Poplin, HafenCity University of Hamburg, Germany
Kevin McDougall, University of South Queensland, Australia
Gianluca Miscione, University of Twente/ITC, The Netherlands
Harlan Onsrud, University of Maine at Orono
Sundeep Sahay, University of Oslo, Norway
David Tulloch, Rutgers University, New Brunswick, New Jersey
Danny Vandenbroucke, Katholieke Universiteit Leuven, Belgium
Mireille van Eechoud, University of Amsterdam, The Netherlands, and Cambridge University, United Kingdom
Jos van Orshoven, Katholieke Universiteit Leuven, Belgium
Jude Wallace, the University of Melbourne, Australia

Introduction

Zorica Nedović-Budić, Joep Crompvoets, and Yola Georgiadou

Motivation for the Book

The coeditors of the book invite you to an international tour of cutting-edge interdisciplinary information and social science research on the implementation and development of spatial data infrastructures (SDIs). The book complements the existing research publications by focusing on nontechnical aspects, expanding the theoretical base, introducing a range of methods, and incorporating empirical investigation and evidence with pragmatic implications. This reading is intended for the students of SDI set in a variety of contexts—from local to national—and roles: technical, administrative, management, and academic. We also hope that the colleagues from allied disciplines to whom the authors of the chapters are connecting will find the book informative and useful in furthering the common theories, scientific agenda, and practice in developing information systems and infrastructures.

SDI—Status Quo

SDIs emerged in the early 1990s as advancements in geospatial and communication technologies (Internet in particular) moved the emphasis from stand-alone geographic information systems (GISs) toward networked and collaborative systems and information infrastructures. United States Executive Order 12906, entitled "Coordinating Geographic Data Acquisition and Access: The National Spatial Data Infrastructure (NSDI)," signed on April 11, 1994 (FGDC 2010); establishing an infrastructure for spatial information in Europe (INSPIRE), adopted by the European Commission in 2007 (CEC 2007); and the longstanding leadership of the Australia–New Zealand Spatial Information Council (ANZLIC 2010) are probably the main conceptual and practical pillars of SDI developments around the globe.

Among many definitions offered over the past two decades, we select Masser's (2005) as comprehensive and clear in stating that spatial data infrastructure

> supports ready access to geographic information. This is achieved through the coordinated actions of nations and organizations that promote awareness and implementation of complementary policies,

> common standards and effective mechanism for the development and availability of interoperable digital geographic data and technologies to support decision making at all scales for multiple purposes. These actions encompass the policies, organizational remits, data, technologies, standards, delivery mechanisms, and financial and human resources necessary to ensure that those working at the (national) and regional scale are not impeded in meeting their objectives. (p. 16)

The ultimate objectives of the SDI initiatives are to promote economic development, to stimulate better cooperation and government, and to foster environmental sustainability. The need for such interventions and improvements in those areas is evident in both developing (GUO 2003) and developed (Sawhill 2002) countries. SDIs are being established across the globe in the most diverse political, institutional, and legal settings, with diverse cultural values, social norms, and levels of economic and technological development (Masser 2005). Since 1994, when the first U.S. Federal Geographic Data Committee's (FGDC) Clearinghouse was established, the number of countries with national spatial data clearinghouses as a key feature of national SDI has been steadily increasing. Masser (2005) identifies 11 countries as SDI innovators and early adopters: Australia, Canada, Indonesia, Japan, Korea, Malaysia, The Netherlands, Portugal, Qatar, United Kingdom, and United States. It is a diverse set, with the majority located in the North and all relatively wealthy societies.

The latest estimates as of July 2010 suggest that about 105 countries have established a national spatial clearinghouse—an almost 25% increase in the past 5 years. The legacy from the most recent available survey conducted in 2005 is considerable regional variability (Crompvoets and Bregt 2007). More than 60% of the countries in Europe and the Americas had established national clearinghouses, whereas less than 20% of African countries had such clearinghouses. The many project initiatives in Africa have been promising, but by 2005 about one quarter of the countries throughout the world had not initiated any plan for such a national facility, likely due to low economic affluence. In Africa,

> SDI capacity building efforts largely have centered on informing colleagues about SDI initiatives, in Africa and elsewhere. This gives countries a smorgasbord of options and leaves them with the challenge to select and adapt one of the documented models to their own environment. Most countries have found it difficult to obtain political support, maintain momentum for SDI development, and [the United Nations Economic Commission for Africa] has limited means to monitor and assess countries' progress. (Lance and Bassolé 2006, p. 335)

Within Europe, INSPIRE State of Play (2010) reports a detailed survey of the directive's transposition, coordination, funding, and sharing measures (Crompvoets and Vandenbroucke 2010). As of February 2010, out of 24 EU member states that responded to the survey, 5 were in the transposition phase* of draft text, 5 with final text, 3 with voted text, and 11 with final text published. Furthermore, it appeared that 13 had a geo-portal and an additional 2 were in the prototyping phase. All this indicates a high variability in the status of INSPIRE transposition across Europe. The main reported obstacles to furthering European national SDIs include ineffective coordination, lack of transposition law, unclear implementation rules, and faltering institutionalization; problems with SDI awareness, finance, data sharing policies, and legislation are also mentioned. Conversely, SDI awareness, presence of an INSPIRE coordinating body, transposition in national legislation, capacity building, establishment of the national geo-portal, and higher support for e-government are listed as the key success factors that INSPIRE has achieved so far.

Research and Challenges

Alongside SDI development activities, the research community has attempted to follow the trends in practice and understand the processes and factors associated with the SDI phenomenon. Substantial research has been undertaken in the technical areas of standardization, networking, access (geo-portals), transfer, and representation of spatial data; policy, organizational, and implementation issues—data sharing and coordination in particular; and SDI evaluation on the nontechnical side. However, developing an SDI and claiming its utility happen under distinctive circumstances that challenge the generalizability and applicability of mostly generic SDI research.

There are stark differences between the North and South, developed and developing countries, industrialized Western economies and economies in transition, and overurbanized and less populated regions with regard to the overall human condition, needs, and the scale of various societal issues. In the case of most impoverished, overpopulated urban regions that lack basic sanitary infrastructure and housing, some would debate even the assumption that information systems and infrastructure are of value or have priority over resolving basic needs (Cavrić, Nedović-Budić, and Ikgopoleng 2003).

While SDI research is maturing and gradually addressing those needs, four limitations can be noted: North centrism, dominance of the national level, technical focus and paucity of theory, and lack of methodological diversity and rigor.

* The transposition phase involves the adoption of national legislation to meet the objectives set by the directive.

North Centrism

The research focus has been primarily on Western and industrialized settings. Research attention to transitional economies in Eastern Europe and the global south is too low (Georgiadou, Puri, and Sahay 2005; de Vries 2006). These regions lag in both SDI establishment and research attention. The north-centric focus in published research contradicts the official rhetoric of the global nature of the geographic information science discipline. Also, SDIs are closely linked to broader social change and, in the context of international development, they are promoted as being important components of processes for poverty reduction and better governance (UNECA 2003, 2005). In that context, we would expect to see strong links between the considerable literature on information systems and international development as well as the articulation of SDIs as critical informational infrastructure.

Sociotechnical themes that are well established in this literature range from the "design–reality gap" conceptual framework—suggested by Heeks (2002) to explore possible reasons for failure of e-governance and IS implementation in developing countries—to the construction of systems that leave little scope for interaction with the host of less formal systems already existing in organizations (Galliers 2003). The design–reality gap is evident in the uneven spatial and economic performance of information systems and infrastructures, and the political goals of inclusion and evenness are seldom matched: The generic "digital divide" encapsulates a variety of spatial and thematic exclusions (Warschauer 2003; Gurstein 2004). A design–reality gap can become also a technocultural gap where IT is implemented without reference to cultural and social conditions (Sahay 1998).

Dominance of the National Level

Researchers' preference has been for national rather than subnational studies, despite the substantive relevance of regional and local levels (Rajabifard et al. 2006). Georgiadou and Blakemore's (2006) bibliometric analysis of a sample of 230 contributions to the Global Spatial Data Infrastructure (GSDI) Association's conferences identifies about half of the papers with a national focus, 24% global, 12% regional, 6% local or provincial each, and only 3% cross-jurisdictional. In particular, cities and urbanized regions are the loci of major societal problems related to poverty, environmental degradation, and overall living condition (Boos and Mueller 2009). Many of them are heavily populated and concentrated areas and the key instances of planning and policy decisions and actions.

SDIs have emerged as a significant area of development in geographic information (GI), where GI forms an underpinning base for wider government strategies and initiatives such as public administration and e-governance. SDIs necessitate a stronger coordination of policy and information landscape—for example, through regional SDI initiatives in Europe (Craglia and Campagna 2009); through the U.S. National Geospatial Programs Office, with a specific societal and governance commitment that states' "current and accurate geospatial data will be available to contribute locally, nationally, and globally to economic growth, environmental quality and stability, and social progress" (NGPO 2005); through the mapping policy in India (Puri 2006); in the context of developing nations' social and economic strategy, such as in Nigeria (Agbaje and Akinyede 2005); and, indeed, at a global but initially limited thematic level with developments such as Google maps (Google 2005) and United Nations Spatial Data Infrastructure (UNSDI) (UNGIWG 2008).

Technical Focus and Paucity of Theory

The majority of research efforts is concentrated on the engineering challenges of SDI development—the endeavor focused at finding out which technological design works in specific administrative circumstances. Georgiadou and Blakemore (2006) identify 12 topics from papers in seven major academic GIS journals published from the late 1990s up to early 2005. While the results do not specifically refer to SDI, they are indicative of the sporadic nature of nontechnical topics such as impact, use, and social issues, represented by only 6, 8, and 5% of the articles, respectively. Clearly, there is paucity of research that systematically analyzes the link between administrative and technological design, on the one hand, and performance (efficiency, administrative transparency, sustainable development, or any other criterion) on the other hand.

More emphasis is needed on the interplay between social (humans, organizations, institutions) and technical issues and the fusion of theoretical perspectives from the social and technical sciences (de Man 2006; Crompvoets, Bouckaert, et al. 2008; Georgiadou and Stoter 2010). The theory of "information infrastructures" (Hanseth and Monteiro 1998) provides potentially a rich basis to develop a deeper theoretical understanding and for deriving sound and practical implications for planning and implementing SDIs (Georgiadou et al. 2005). It has its roots in the sociology of technology research tradition (Callon and Law 1986; Latour 1987) and has been extensively applied in IS research to analyze networked systems whose development is controlled by multiple actors. The information infrastructure perspective emphasizes that the social and technical are not separable and are instead constituted by and constitutive of one another (Latour 1999).

In addition to connecting to the information infrastructure theory and approach, Budhathoki and Nedović-Budić (2007) suggest additional sources to be consulted and incorporated in the SDI research base (e.g., interorganizational collaboration–cooperation–coordination (3Cs), intergovernmental relations, actor network theory (ANT), and use–utility–usability of information systems, as well as policy implementation, development of federated databases and systems, capacity building, and public administration and finance. Echoing the concerns of GIS and SDI researchers, public administration scholars warn against unidirectional causal relationships between ICT and governance and speak on purpose about "implications" because "autonomous political, legal, economic and professional developments in and around public administration, and the changes in ideas and ideals for that matter, are as important for the effects of ICT applications on public administration as the technological developments themselves" (van de Donk and Snellen 2002, p. 11).

The list of sources is not exhaustive; in fact, none of the disciplines individually offers a comprehensive knowledge base required to develop and sustain SDI networks. The interdisciplinary view and approach are necessary for understanding and developing SDIs.

Lack of Methodological Diversity and Rigor

Research frameworks and methodologies do not yet adequately reflect the interdisciplinary nature of the SDI phenomenon. A few scholars have focused on theoretically grounding their empirical work (Budhathoki and Nedović-Budić 2007; Onsrud 2007) but, generally, the investigations have been mainly technoscientific and monodisciplinary. Opposed to the rigorous scientific research, the SDI work is probably more present in the "gray" literature (e.g., in conference proceedings) than in academically referenced sources. Georgiadou and Blakemore (2006) find in the same sample of GSDI conference proceedings mentioned earlier that over 95% of the papers use a positivist paradigm with only 3 and 2% of theoretically grounded and critical research, respectively; 61% of the papers are represented by a project report, review, or future plan; 25% are based on opinion, anecdotes, or visions; and only 10% apply a scientific method: survey (7%) and case study (3%).

As Wilson (2000) points out, the evolution of technological systems is necessarily supported by rigorous empirical research conducted using multidisciplinary perspectives. Given the nature of the SDI phenomenon, research teams should be constituted in interdisciplinary terms (sociologists, anthropologists, geoscientists, information-system researchers, and economists) and supported to conduct longitudinal research (rather than cross-sectional studies) that can follow the unfolding of the SDI dynamics over time. Also, as IS research has emphasized, implementation analysis is well guided by an interpretive epistemology, where the different social meanings constructed by various stakeholder groups are identified, as a complement to a positivist approach that assumes objectivity of data and statistical generalizations (Walsham 1993; Klein and Myers 1999). The perceived challenge in current SDI research relates to conducting independent, verifiable, and repeatable studies with systematically acquired hard (as opposed to anecdotal) evidence, based on a variety of quantitative and qualitative methods.

Objective

The objective of this book is to try to redress some of these limitations, advance the scientific discourse, and contribute to SDI practice. We sought theoretically and empirically sound SDI research to contribute to better informed and more useful SDI developments, including the regional and local levels in the North and the South. The idea for the book was conceived during a preconference workshop of AGILE 2008 in Girona (Spain), a PhD workshop held at Delft University of Technology in summer 2009 that preceded the combined 11th Global Spatial Data Infrastructure Association, 3rd INSPIRE, and Geonovum-Space for Geo-Information conference. The majority of the contributions are from PhD students or recent graduates who are on the cutting edge of the new interdisciplinary SDI scientific ideas and research trends. The conducted research is based at the Katholieke Universiteit Leuven (KUL), the Faculty for Geo-Information Science and Earth Observation (ITC) at the University of Twente, Delft University of Technology, and the University of Illinois at Urbana-Champaign (UIUC).

The book's chapters explore SDIs in local, regional, national, and international contexts; the different forms they take; and the factors that influence their establishment and use under different circumstances. The presented work is contextually embedded and suggests a comparative perspective, although it does not allow for a direct comparison of the cases. The research was designed and pursued independently by several groups and individuals, and no common or single conceptual framework of methodological approach was used. The presented work builds on a variety of disciplinary knowledge and theoretical frameworks and includes empirical evidence based on primary data collection. Ten research chapters are followed by a commentary in which the contributions are discussed and connected to the relevant scientific discourses.

Previous Research Output

The publication output in the area of SDI is not extensive, although it is significant in breaking new ground in research and societal practice. Groot and McLaughlin's *Geospatial Data Infrastructure: Concepts, Cases and Good Practice* (2000); Masser's *GIS Worlds—Creating Spatial Data Infrastructures* (2005) and *Building European Spatial Data Infrastructure* (2007); Nebert's *SDI Cookbook* (2004); and Williamson, Rajabifard, and Feeney's *Developing Spatial Data Infrastructures—From Concepts to Reality* (2003) discuss the SDI phenomenon as it emerges and diffuses within national settings and internationally, define the main related concepts, and present case studies and trends. As in van Loenen and Kok's *Spatial Data Infrastructure and Policy Development in Europe and the United States* (2004), the national level is most commonly featured.

Onsrud's book, *Research and Theory in Advancing Spatial Data Infrastructure Concepts* (2007), delineates SDI theory and contributes not only by describing existing practices, but also by exploring the relationships between societal and institutional forces and information technologies and infrastructures in general. The book is eclectic in its thematic coverage and with variable presence of theoretical grounding. Complementing those works with an evaluative perspective is the volume *A Multi-View Framework to Assess Spatial Data Infrastructures* (2008) by Crompvoets, Rajabifard, et al.

Some of the other important and valuable texts, like Moellering's *World Spatial Metadata Standards* (2005); Burkholder's *The 3-D Global Spatial Data Model: Foundation of the Spatial Data Infrastructure* (2008); Nogueras-Iso, Javier Zarazaga-Soria, and Muro-Medrano's *Geographic Information Metadata for Spatial Data Infrastructures: Resources, Interoperability and Information Retrieval* (2005); and van Oosterom and Zlatanova's *Creating SDIs—Toward the Spatial Semantic Web* (2008), are highly specialized in technical standardization, domain modeling, and ontological aspects of SDI. Some are practice oriented, such as Sadahiro's *Spatial Data Infrastructure for Urban Regeneration* (2008) and van Loenen, Besemer, and Zevenbergen's *SDI Convergence, Research, Emerging Trends, and Critical Assessment* (2009), which deals with legal, funding, catalogues, domain modeling, metadata, evaluation, application, and governance topics surrounding service-oriented SDIs and prototype geoportals. Finally, there are publications in languages other than English—for example, in Spanish, *Avances en las Infraestructuras de Datos Espaciales* by Granell and Gould (2006), *Infraestructuras de Datos Espaciales en Iberoamérica y el Caribe* by Delgado Fernández and Crompvoets (2007), and *Construyendo Infraestructuras de Datos Espaciales a Nivel Local* by Delgado Fernández and Cruz Iglesias (2009); and, in German, *Geodaten-Infrastruktur, Grundlagen und Anwendungen* by Bernard, Fitzke, and Wagner (2005).

More rigorous SDI research has been disseminated in various journals (mostly in the areas of geospatial science and technology) such as the *International Journal of Geographical Information Science* and the *Journal of the Urban and Regional Information Systems Association.* Some are published in interdisciplinary or other disciplinary journals, such as the special 2006–2007 double issue of the international journal, *Information Technology for Development* (ITD), and some of the urban planning journals (e.g., *Environment and Planning B* and *Computers, Environment and Urban Systems*). The double special issue of ITD drew together researchers from the information infrastructure and spatial data infrastructure research communities with the common interest of understanding implementation of information infrastructures in developing countries (Georgiadou, Bernard, and Sahay 2006, 2007).

Last but not least, the *International Journal of Spatial Data Infrastructure Research,* published under the sponsorship of the European Commission's Joint Research Center and specifically focused on SDI-related topics, has provided an invaluable service to both the academic and professional communities. Since its inception in 2006, IJSDIR has become a venue for SDI-related articles from a variety of disciplinary and regional perspectives and also introduced innovations like volunteered geographic information (VGI) and digital earth and global earth observation system of systems (GEOSS), raised awareness about SDI development and policies in Europe (e.g., Norway, England, Croatia) and other continents (e.g., India), emphasized the importance of evaluating SDI through methodologically sound measurement of impact, and presented a variety of technical topics related to standardization and domain data modeling.

While the books and articles in these journals have an international reach and contribute to the progression of the SDI field and geospatial science, research output that explicitly requires empirical testing of the inter- and multidisciplinary theoretical approaches and systematically addresses the contextual elements is still scarce. The difficulty in making substantial progress is also due to the scattered nature of SDI research conducted by individual scientists. In this dispersed landscape, several project nodes promise to advance the knowledge about SDI considerably:

- *Space for Geo-Information* (The Netherlands). The research program aims to improve and innovate the national geo-information infrastructure and the field of geospatial knowledge in The Netherlands for satisfactory and efficient administration and powerful industry (http://www.rgi.nl/).
- *Cooperative Research Center (CRC) for Spatial Information* (Australia). The program defines research as the innovative use and spatial application of emerging geo-technologies, as well as the development of new technologies (http://www.spatialinfocrc.org/).
- *GEOIDE, Geomatics for Informed Decisions* (Canada). The program aims to consolidate and strengthen the Canadian geomatics industry, while making optimum use of Canada's research and development resources (http://www.geoide.ulaval.ca/home.aspx).
- *Joint Research Center, Institute for Environment and Sustainability* (European Commission). The institute is responsible for instituting the European INSPIRE Directive and initiating pertinent policy and evaluation research on SDI practices across the European Union (http://ies.jrc.ec.europa.eu/).

Examples of smaller SDI projects include:

- *SPATIALIST, Spatial Data Infrastructure and Public Sector Innovation* (Belgium). The program aims to determine the technical, legal, economic, sociological, and public administrative requirements to further develop a spatial data infrastructure in Flanders that is consistent with international standards and efficient, effective, flexible, and feasible (www.spatialist.be).

- *NWO-Wotro "Using Spatial Information Infrastructure in Urban Governance Networks: Reducing Urban Deprivations in Indian Cities"* (The Netherlands). The research program focuses on how urban governance networks can tackle urban inequalities and household deprivations in large Indian cities by using local spatial information infrastructure (http://www.itc.nl/research/themes/infgovn/projects/spatial_information_infrastructure_in_urban_governance_networks.asp).

These are complemented by scattered research efforts as part of dissertations or small-grant research—for example, dissertations by Kate T. Lance and El-Sayed Omran (University of Wageningen/ITC), Kevin McDougall (University of Melbourne), Elisabetta Genovese (Laval University), Nama Raj Budhathoki (University of Illinois at Urbana-Champaign), and Ruben Béjar (University of Zaragoza). The lists provided here do not claim to be exhaustive, but rather are illustrative of the relatively small output and community gathered around research on SDI implementation and institutionalization.

Book Organization and Contents

The volume delivered here cannot respond to all demands of SDI research and practice mentioned in this introduction, but it attempts to provide examples of research efforts that would raise their quality in terms of theory-based, empirically grounded research with an interdisciplinary and international draw and reach, as well as implication for SDI practice. Seven of the 11 chapters included in this volume represent research conducted by PhD students or recent graduates. Three contributions are based on the spatial data infrastructure and public sector innovation (SPATIALIST, KUL, Belgium) project, two on ITC projects including the NWO-Wotro using spatial information infrastructure in urban governance networks: reducing urban deprivations in Indian cities (ITC) and ITC staff research in Uganda; one on Dutch space for a geo-information program with a project on geo-portals; and one on exploration of motivations for VGI conducted at the University of Illinois (UIUC). The contributions have gone through a full, double-blind peer review process.

An additional three chapters are reprints of previously published peer-reviewed journal articles: one from the *Journal of Urban and Regional Information Systems Association* and two from the special issue on implementation of spatial data infrastructures in transitional economies of the international journal, *Information Technology for Development*. The coeditors had to resort to this option because of the difficulty of identifying and securing contributions that would meet the requirements regarding theory-based research with rigorous methodology and empirical work. The final chapter presents a commentary on the studies included in the book and connects them to the scientific discourse on information science and infrastructures.

The book is organized into four parts: I. Institutions and Organizations; II. Data and Technology; III. People and Practices; and IV. Sense-Making and Summing Up—and 11 chapters.

Part I: Institutions and Organizations

We start the book with Part I on institutions and organizations as the most rigid element of any project implementation effort, but indeed crucial to understand, adapt, or change. Such a system of social, political, and economic norms and rules translated into administrative mechanisms and organizational bodies that enforce them is one of the most influential constituents in the development and use of SDIs at all levels. The four chapters in this section refer to land administration, structure of public organizations, interorganizational relationships, and legal definition of a public task regarding distribution of spatial data at the national (Guatemala), subnational (Flanders), local (United States), and international (European Union) levels, respectively. The discussion in these chapters revolves around relationships and issues of institutional and organizational cooperation, forms, mechanisms, and frameworks that could impede or facilitate establishment of SDIs.

The first contribution by Janssen, Crompvoets, and Dumortier, takes on the public sector as increasingly confronted with demands to make its spatial data available but with unclear scope of its role and *public task* in providing such data. The authors look at the EU's public sector information (PSI) and infrastructure for spatial information in Europe (INSPIRE) directives and argue that criteria such as legal basis, link to the public body's core responsibility, market failure, public interest, and public goods are not sufficient to determine the public task of the public sector to provide spatial data or spatial information services. They suggest that, while defining the public task may not be possible, its uncertain character could be reduced through participatory democratic processes involving a broad range of interested stakeholders.

Silva's chapter on Guatemala uses a critical hermeneutic analysis, sociology of translation, and actor-network theory (ANT) approaches to explore interinstitutional dynamics associated with the post-civil-war land administration reform, the power struggle it triggered, and the clashes between the Guatemalan and European views on the ownership of private property. The study is deeply embedded in the local context and history and concludes with an implementation model that may suggest ways of overcoming the political aspects and irrationalities of institutionalization.

In the next chapter, Dessers, Van Hootegem, Crompvoets, and Hendriks argue that the analysis of spatial information flows should not be separated from the business processes in which they are embedded. The performance of both is expected to be influenced by the structural characteristics of the organizations involved. Based on theoretical considerations, functional concentration is proposed as a central concept for describing these organizational structures. The ability of this concept to discern between different forms of organizational structures is explored through a case study of the Flanders region in Belgium.

The final chapter in Part I, by Nedović-Budić, Pinto, and Warnecke, examines various properties of data sharing activities, as well as related motivations for inter- and intraorganizational cooperative relationships. The authors draw from the literature on interorganizational relationships and report results of a national survey. The reasons and nature of cooperation they isolate for external interactions are most relevant for the setups and dynamics expected in SDIs. They find that interorganizational relationships tend to be driven by common missions/goals and financial resources, guided by formalized mechanisms, and inclined toward less complex interactions. Even almost a decade after the original data collection, this research carries relevance and implications for future efforts to induce wider sharing of geographic information across organizational boundaries and to build SDIs at all levels. On the quantitative–qualitative spectrum of methodologies, it represents the quantitative extreme.

Part II: Data and Technology

Part II is about data and technology—two core elements of any information system or infrastructure that tend to preoccupy the attention of both SDI practitioners and scholars. This section looks at them through an institutional and organizational lens because both data and technology interact with or are subjected to organizational processes. The authors of the three chapters included in this part observe these overemphasized and time-consuming aspects of SDIs in the context of Uganda, Flanders, and The Netherlands.

De Vries and Lance analyze public sector practices in Uganda using resource dependency theory (RDT) and an axial coding method to consider specifically how the handling of power differences and uncertainty explain existing redundancies and efficiencies. The analysis shows that organizations tend to prioritize their single mandates over cross-organizational efficiencies in view of a possible power loss and emerging uncertainties. Furthermore, the presence of a large number of donor-funded projects contributes to both redundancies and heterogeneities in data quality.

Vancauwenberghe, Crompvoets, Bouckaert, and Vandenbroucke rely on social network analysis to define and analyze SDIs as dynamic and heterogeneous interactions between a large number of spatial data users, producers, and suppliers. In the network perspective, an SDI is operationalized in terms of the organizations producing and using spatial data in a shared environment and the flows of spatial data between these organizations. The authors examine the specific role and position of different actors and arrangements within this network in Flanders, Belgium. The analysis indicates that almost half of the spatial data exchanges are a direct result of the partnership "SDI-Flanders"—the central SDI arrangement in Flanders. They find that access to and exchange of spatial data in Flanders are facilitated also by several other arrangements. The chapter evaluates how social network analysis can be used as a methodological approach to analyze the complex interactions between different actors and arrangements within a network of spatial data exchanges.

In the third contribution to Part II, Koerten and Veenswijk use a narrative approach to elicit sense-making processes in order to get a better understanding of the development of the Dutch geo-portal project. Intensive research was carried out by observing project meetings and conferences and interviewing key persons, both within and outside the project environment. The authors' ethnography records day-to-day struggles with project goals, technology, and infrastructure. They found that project participants cannot easily distinguish between requirements for infrastructure and innovation. While infrastructures need stable environments with harsh standardization that will last, innovation challenges toward new developments and uncontested terrains. They argue that this discrepancy is a cause for serious redefinitions of NGII project goals, assessment rules, and results.

Part III: People and Practices

The ultimate expectation of the geospatial technologies introduced over the past three decades has been the broad societal benefit in terms of democracy and local empowerment in decision making. The SDI developments, however, still tend to be detached from local needs and processes and do not connect well to the citizen. Part III addresses this deficiency in SDI literature and practice with three contributions: one on the local actors and practices in the city of Mugdali, India; one on the empowerment brought about through SDI implementation driven by local applications (phronesis) in the Brazilian city of Belo Horizonte; and one on the emergence of the new data access and contribution technologies and mechanisms—VGI, geo-browsers, and c sensors.

In the first chapter, Richter, Miscione, De', and Pfeffer draw on empirical data from 7 months of qualitative research in the southern Indian city of Mugdali to analyze local practices in the case of the slum declaration process. They conceptualize this process as the interplay of two practices: classificatory and nonclassificatory listings. They use the term "listings" in order to make explicit what people are actively doing (listing)—the practices through which lists of slum areas, inhabitants, and boundary drawings are created and used. In the empirical case of slum declaration, classificatory listings are driven by public administration procedures, whereas nonclassificatory listings are more situation dependent and driven by changing local sociopolitical and sociospatial relations. Taking into consideration the potential shifts in practices and actors' roles that SDI implementation in Mugdali would require, the authors discuss "infrastructuring" as an alternative implementation strategy, where boundaries between design(er) and use(r) become blurred to give more scope to actors in shaping their own role vis-à-vis geographic information. This chapter provides probably the best illustration of how strikingly complex and different the local practices in the South could be from the northern notions of local government activities with regard to substandard settlements and structures.

In the second chapter of this section, Davis and Fonseca use Gadamer's concept of phronesis to show how an application-driven project is a key to success. The authors also draw on Habermas's ideas to show the importance of emancipatory knowledge in the implementation of SDIs. The authors present a case study on a GIS project in Belo Horizonte, Brazil. The project has been evolving for 15 years; it started with a focus on data and standards and generated a strong and active spatial data infrastructure for the city. The reasons for success were many, including the application-driven nature of the project along with the combination of multiple disciplines and multiple levels of expertise in its design and implementation team.

Part III concludes with Georgiadou, Budhathoki, and Nedović-Budić's review of the intensive evolution of new means of data access and production that may be more suitable to the technological capacity and social and institutional relations in the context of developing countries. In Africa, SDI development is still fraught with problems. Communication via mobile phones and occasional access to the Internet are stimulating the emergence of bottom-up infrastructures based on VGI, geo-browsers, and citizen-sensors. The authors discuss two partially overlapping developments: (a) more mature VGI by participants with different levels of spatial expertise, and (b) a more tentative development of citizen-sensor networks—at its simplest, a combination of citizens texting grievances through mobile phones and Web map services that publicize these grievances on the Web. Citizens as "citoyens" are enabled to participate and influence local decisions and services and to capture their constituent power.

Part IV: Sense-Making and Summing Up

The book concludes with a commentary chapter by Miscione and Vandenbroucke, who deal with the material presented in the previous 10 chapters with a view on the differences and commonalities between North and South, unit of analysis (local, regional, national), sociotechnical theoretical base, and explanatory versus prescriptive research and methodologies. The authors connect the research included in this volume to the concepts, frameworks, and debates on information infrastructures, science, and epistemology. They compare and contrast the contributions from the North and South and speculate on the issues brought about by the diversity of contexts and needs and on the ways it can be addressed by science and practice.

Following this commentary, the coeditors close the volume with their summary of contributions, the key findings, and implications for future SDI research and practice. They acknowledge that the path is long and there are many limitations and obstacles ahead. This volume redresses some, but not all, despite the ambitious aspirations. However, the book's guiding principle of encouraging and promoting explanatory inter- and multidisciplinary SDI research is followed strictly, along with ensuring the empirical and pragmatic edge. Sociotechnical approaches, the user perspective, and the need for comparative studies that point to the key system-related and contextual determinants of the SDI phenomenon also are entrenched in most contributions. Metaphorically, our message is well expressed by Edwards et al. (2007):

> Moving between social organization and technical infrastructure is like crossing the Northwest Passage: seasonal shifts in ice mean that the voyage can be made, but never in the same way twice. Under such conditions, what is needed are not rigid maps, but flexible and creative principles of navigation. (p. 33)

Let the journey begin!

References

Agbaje, G. I., and J. O. Akinyede. 2005. NGDI development in Nigeria: Policy issues on information access and information dissemination. April 23–28, report, 13 pp. Addis Ababa: United Nations Economic Commission for Africa, Committee on Development Information.

ANZLIC (Australia New Zealand Land Information Council). 2010. The Spatial Information Council. http://www.anzlic.org.au/ (accessed July 29, 2010).

Bernard, L., J. Fitzke, and R. M. Wagner. 2005. *Geodaten-Infrastruktur, Grundlagen und Anwendungen.* Heidelberg: Herbert Wichmann Verlag (in German).

Boos, S., and H. Mueller. 2009. Evaluation of spatial information technology applications for mega city management. In *SDI convergence, research, emerging trends and critical assessment,* ed. B. van Loenen, J. W. J. Besemer, and J. A. Zevenbergen, 48, 189–203. Delft: Netherlands Geodetic Commission.

Budhathoki, N. R., and Z. Nedović-Budić. 2007. Expanding the spatial data infrastructure knowledge base. In *Research and theory in advancing spatial data infrastructure concepts,* ed. O. Harlan, 7–31. Redlands, CA: ESRI Press.

Burkholder, E. R. 2008. *The 3-D global spatial data model: Foundation of the spatial data infrastructure.* Boca Raton, FL: CRC Press.

Callon, M., and J. Law, eds. 1986. *Mapping the dynamics of science and technology. Sociology of science in the real world.* London: The Macmillan Press Ltd.

Cavrić, B. I., Z. Nedović-Budić, and H. G. Ikgopoleng. 2003. Diffusion of GIS technology in Botswana: Process and determinants. *International Development Planning Review* 25:195–219.

CEC (Commission of the European Communities). 2007. Directive 2007/2/EC of the European Parliament and of the Council of 14 March 2007 establishing an infrastructure for spatial information in the European Community (INSPIRE). http://eur-lex.europa.eu/JOHtml.do?uri=OJ:L:2007:108:SOM:EN:HTML (accessed July 29, 2010).

Craglia, M., and M. Campagna, 2009. Advanced regional spatial data infrastructures in Europe. Workshop report. Ispra, Italy: Joint Research Centre Institute for Environment and Sustainability.

Crompvoets, J., G. Bouckaert, G. Vancauwenberghe, et al. 2008. Interdisciplinary research project: SPATIALIST; spatial data infrastructures and public sector innovation in Flanders (Belgium). In *Proceedings of GSDI-10 Conference, Small Island Perspectives on Global Challenges: The Role of Spatial Data in Supporting a Sustainable Future*. St. Augustine, Trinidad. Abstract and paper, 24 pp.

Crompvoets, J., and A. Bregt. 2007. Worldwide development of national spatial data clearinghouses (2000–2005). In *Research and theory in advancing spatial data infrastructure concepts*, ed. H. Onsrud, 133–145. Redlands, CA: ESRI Press.

Crompvoets, J., A. Rajabifard, B. van Loenen, and T. Delgado Fernández, eds. 2008. *A multi-view framework to assess spatial data infrastructures*. Wageningen, The Netherlands: Space for Geo-Information (RGI), Wageningen University; Melbourne, Australia: Centre for SDIs and Land Administration, Department of Geomatics, University of Melbourne.

Crompvoets, J., and D. Vandenbroucke. 2010. INSPIRE & NSDI state of play, D3.1. Detailed survey on coordination, funding and sharing measures. Leuven, Belgium: Katholieke Universiteit Leuven.

de Man, W. H. E. 2006. Understanding SDI: Complexity and institutionalization. *International Journal of Geographical Information Science (IJGIS)* 20:329–343.

de Vries, W. T. 2006. Why local spatial data infrastructures SDIs are not just mirror reflections of national SDI objectives: Case study of Bekasi, Indonesia. *The Electronic Journal on Information Systems in Developing Countries* (EJISDE) 27(2006) art4, 28 pp. http://www.ejisdc.org/ojs2/index.php/ejisdc/article/view/267

Edwards, P. N., S. J. Jackson, G. C. Bowker, et al. 2007. Report of Workshop "History & Theory of Infrastructure: Lessons for New Scientific Cyberinfrastructures." NSF Grant 0630263, Human and Social Dynamics, Computer and Information Science and Engineering, Office of Cyberinfrastructure, January 2007. http://deepblue.lib.umich.edu/bitstream/2027.42/49353/3/UnderstandingInfrastructure2007.pdf

Delgado Fernández, T., and J. Crompvoets. 2007. *Infraestructuras de datos espaciales en Iberoamérica y el Caribe*. Havana, Cuba: IDICT (in Spanish).

Delgado Fernández, T., and R. Cruz Iglesias. 2009. *Construyendo Infraestructuras de datos espaciales a nivel local*. Havana, Cuba: CUJAE (in Spanish).

FGDC (Federal Geographic Data Committee). 2010. Executive order 12906: Coordinating geographic data acquisition and access: the national spatial data infrastructure, signed by President Bill Clinton on April 11, 1994. http://www.fgdc.gov/nsdi/policyandplanning/executive_order (accessed July 20, 2010).

Galliers, R. D. 2003. Change as change as crisis or growth? Towards a transdisciplinary view of information systems field of study. Response to Benbasat and Zmud's call for returning to the IT artifact. *Journal of the Association for Information Systems* 4:337–351.

Georgiadou, Y., L. Bernard, and S. Sahay. 2006. Implementation of spatial data infrastructures in transitional economies: Editorial introduction to part one of the special issue. *Information Technology for Development* 12:247–253.

Georgiadou, Y., L. Bernard, and S. Sahay. 2007. Implementation of spatial data infrastructures in transitional economies: Editorial introduction to part two of the special issue. *Information Technology for Development* 13:1–5.

Georgiadou, Y., and M. Blakemore. 2006. A journey through GIS discourses. Unpublished paper. Enchede, The Netherlands: ITC.

Georgiadou, Y., S. K. Puri, and S. Sahay. 2005. Towards a potential research agenda to guide the implementation of spatial data infrastructures: A case study from India. *International Journal of Geographical Information Science* 19:1113–1130.

Georgiadou, Y., and J. E. Stoter. 2010. Studying the use of geo-information in government: A conceptual framework. *Computers, Environment and Urban Systems* 34:70–78.

GUO (Global Urban Observatory). 2003. *Slums of the world: The face of urban poverty in the new millennium?* Nairobi, Kenya: United Nations Human Settlements Program (UN-HABITAT).

Google. 2005. Google maps. http://maps.google.com/ (accessed February 10, 2005).

Granell, C., and M. Gould, eds. 2006. Avances en las infraestructuras de datos espaciales. *Collecció treballs d'informàtica i tecnologia,* 26 Castelló de la Plana, Spain: Publications de la Universitat Jaume I (in Spanish).

Groot, R., and J. McLaughlin, eds. 2000. *Geospatial data infrastructure: Concepts, cases and good practice.* New York: Oxford University Press, Inc.

Gurstein, M. 2004. Effective use and the community informatics sector: Some thoughts on Canada's approach to community technology/community access. *Communications in the Public Interest* 2: 223–244. Ottawa: Canadian Centre for Policy Alternatives.

Hanseth, O., and E. Monteiro. 1998. Understanding information infrastructure. Unpublished manuscript. http://heim.ifi.uio.no/~oleha/Publications/bok.html (accessed August 5, 2010).

Heeks, R. 2002. Information systems and developing countries: Failure, success, and local improvisations. *Information Society* 18:101–112.

Klein, H. K., and M. D. Myers. 1999. A set of principles for conducting and evaluating interpretive field studies in information systems. *MIS Quarterly* 23:67–93.

Lance, K. T., and A. Bassolé. 2006. SDI and national information and communication infrastructure NICI integration in Africa. *Information Technology for Development* 12:333–338.

Latour, B. 1987. *Science in action: How to follow scientists and engineers through society.* Cambridge, MA: Harvard University Press.

Latour, B. 1999. *Pandora's hope. Essays on the reality of science studies.* Cambridge, MA: Harvard University Press.

Masser, I. 2005. *GIS worlds: Creating spatial data infrastructures.* Redlands, CA: ESRI Press.

Masser, I. 2007, *Building European spatial data infrastructures.* Redlands, CA: ESRI Press.

Moellering, H., ed. 2005. *World spatial metadata standards: Scientific and technical characteristics, and full descriptions with cross table.* Oxford, England: Elsevier Science.

NGPO (National Geospatial Programs Office). 2005. United States geological survey. http://www.usgs.gov/ngpo/index.html (accessed April 9, 2005).

Nebert, D. D. 2004. *Developing spatial data infrastructures: The SDI cookbook.* Global Spatial Data Infrastructure (GSDI), version 2.0. GSDI Association. http://www.gsdi.org/docs2004/Cookbook/cookbookV2.0.pdf

Nogueras-Iso, J., F. Javier Zarazaga-Soria, and P. R. Muro-Medrano. 2005. *Geographic information metadata for spatial data infrastructures: Resources, interoperability and information retrieval.* Berlin: Springer–Verlag.

Onsrud, H., ed. 2007. *Research and theory in advancing spatial data infrastructure concepts.* Redlands, CA: ESRI Press.

Puri, S. K. 2006. Technological frames of stakeholders shaping the SDI implementation: A case study from India. *Information Technology for Development* 12:311–331.

Rajabifard, A., A. Binns, I. Masser, and I. P. Williamson. 2006. The role of sub-national government and the private sector in future SDIs. *International Journal of Geographical Information Science* 20:727–741.

Sadahiro, Y. 2008. *Spatial data infrastructure for urban regeneration.* Berlin: Springer–Verlag.

Sahay, S. 1998. Implementing GIS technology in India: Some issues of time and space. *Accounting, Management and Information Technologies* 8:147–188.

Sawhill, I. V. 2002. Poverty in the United States. In *The concise encyclopedia of economics,* ed. D. R. Henderson. Liberty Fund Inc. http://www.econlib.org, http://www.econlib.org/library/Enc/PovertyinAmerica.html (accessed on April 11, 2006).

UNECA (United Nations Economic Commission for Africa). 2003. *SDI implementation guide for Africa.* Endorsed by the Third Meeting of the Committee on Development Information. Addis Ababa, Ethiopia. United Nations Economic Commission for Africa.

UNECA (United Nations Economic Commission for Africa). 2005. Concept paper: CODI IV. Information as an economic resource. Addis Ababa: United Nations Economic Commission for Africa, Committee on Development Information. April 23, report E/ECA/CODI/4/INF/3, 8 pp.

UNGIWG (United Nations Geographic Information Working Group). 2008. Resolution by the 9th Plenary Meeting of the UNGIWG in support of the development of a United Nations spatial data infrastructure (UNSDI). http://www.ungiwg.org/unsdi.htm (accessed July 15, 2010).

van de Donk, W. B. H. J., and I. Th. M. Snellen. 2002. Towards a theory of public administration in an information age? In *Public administration in an information age,* ed. I. Th. M. Snellen and W. B. H. J. van de Donk, 3–19. Amsterdam: IOS Press.

van Loenen, B., J. W. J. Besemer, and J. A. Zevenbergen, eds. 2009. *SDI convergence, research, emerging trends, and critical assessment.* Delft: Netherlands Geodetic Commission.

van Loenen, B., and B. C. Kok, eds. 2004. *Spatial data infrastructure and policy development in Europe and the United States.* Delft, The Netherlands: Delft University Press.

van Oosterom, P., and S. Zlatanova, eds. 2008. *Creating SDIs—Toward the spatial semantic Web.* Boca Raton, FL: CRC Press.

Walsham, G. 1993. *Interpreting information systems in organizations.* Chichester, England: John Wiley & Sons.

Warschauer, M. 2003. *Technology and social inclusion: Rethinking the digital divide.* Cambridge, MA: The MIT Press.

Williamson, I., A. Rajabifard, and M-E. F. Feeney. 2003. *Developing spatial data infrastructures—From concepts to reality*. London: Taylor & Francis.

Wilson, E. J., III. 2000. The what, why, where and how of national information infrastructures. In *National information infrastructure initiatives: Vision and policy design*, ed. B. Kahin and E. J. Wilson, III, 1–23. Cambridge, MA: MIT Press.

Editors

Zorica Nedović-Budić is professor and chair of spatial planning and geographic information systems in the School of Geography, Planning and Environmental Policy at University College Dublin. Her main area of interest is the intersection of information and communication technologies and local planning and governance; the specific focus is on development and implementation of geographic information systems and spatial data infrastructures and evaluation of their impact on the local planning process and decisions. Dr. Nedović-Budić has served on the board of directors of the Urban and Regional Information Systems Association (URISA) and the University Consortium for Geographic Information Science (UCGIS) and as the book reviews coeditor for the *Journal of the American Planning Association*. She is currently an editorial board member of the *URISA Journal, International Journal of Spatial Data Infrastructure, Territorium, and International Journal of Knowledge-Based Development.*

Joep Crompvoets is senior researcher at the Public Management Institute of Katholieke Universiteit Leuven (Belgium) and the Center for Spatial Data Infrastructures and Land Administration of the University of Melbourne (Australia). Since 2007, he has been project coordinator of the interdisciplinary project SPATIALIST: Spatial Data Infrastructures and Public Sector Innovation in Flanders, Belgium. Dr. Crompvoets is a respected expert who is specialized in the development and research of spatial data infrastructures.

Yola Georgiadou is professor in geo-information for governance at the Faculty for Geo-information Science and Earth Observation (ITC), University of Twente, The Netherlands. She served as a member of the board of the GSDI Association (representing academia) and in the Capacity Building Working Group of CODI-Geo, United Nations Economic Commission of Africa. She serves in the Dutch Commission of Geodesy (NCG in Dutch) and the subcommission on SDI in The Netherlands. She is a member of the editorial boards of the *Journal of Information Technology for Development* (JITD), the *International Journal of SDI Research* (IJSDIR), and the *International Journal of Digital Earth* (IJDE). Her research interests include the use of geo-information in public governance and the governance of spatial data infrastructures.

Contributors

Geert Bouckaert
Public Management Institute
Katholieke Universiteit Leuven
Leuven, Belgium

Nama Raj Budhathoki
University of Illinois at Urbana-Champaign
Champaign, Illinois
Department of Geography, McGill University
Montreal, Canada

Joep Crompvoets
Public Management Institute
Katholieke Universiteit Leuven
Leuven, Belgium

Clodoveu A. Davis, Jr.
Informatics Institute
Pontifical Catholic University of Minas Gerais
PUC Minas, Belo Horizonte, Brazil

Rahul De'
Indian Institute of Management Bangalore
Bangalore, India

Ezra Dessers
Center for Sociological Research
Katholieke Universiteit Leuven
Leuven, Belgium

Walter T. de Vries
Faculty of Geo-Information Science and Earth Observation
University of Twente
Enschede, The Netherlands

Jos Dumortier
Interdisciplinary Center for Law and ICT
Katholieke Universiteit Leuven
Leuven, Belgium

Frederico Fonseca
College of Information Sciences and Technology
The Pennsylvania State University
University Park, Pennsylvania

Yola Georgiadou
Faculty of Geo-Information Science and Earth Observation
University of Twente
Enschede, The Netherlands

Paul H. J. Hendriks
Institute for Management Research
Radboud University Nijmegen
Nijmegen, The Netherlands

Katleen Janssen
Interdisciplinary Center for Law and ICT
Katholieke Universiteit Leuven
Leuven, Belgium

Henk Koerten
OTB Research Institute
Delft University of Technology
Delft, The Netherlands

Kate T. Lance
NASA Marshall Space Flight Center
National Space Science and Technology Center (NSSTC)
Huntsville, Alabama

Gianluca Miscione
Faculty of Geo-Information Science and Earth Observation
University of Twente
Enschede, The Netherlands

Zorica Nedović-Budić
School of Geography, Planning and Environmental Policy
University College Dublin
Dublin, Ireland

Harlan Onsrud
Department of Spatial Information Science and Engineering
University of Maine at Orono
Orono, Maine

Karin Pfeffer
Faculty of Social and Behavioral Sciences
University of Amsterdam
Amsterdam, The Netherlands

Jeffrey K. Pinto
Black School of Business
Pennsylvania State University
Erie, Pennsylvania

Christine Richter
Faculty of Geo-Information Science and Earth Observation
University of Twente
Enschede, The Netherlands

Leiser Silva
C. T. Bauer College of Business
Houston, Texas

Glenn Vancauwenberghe
Public Management Institute
Katholieke Universiteit Leuven
Leuven, Belgium

Danny Vandenbroucke
Spatial Applications Division (SADL)
Katholieke Universiteit Leuven
Leuven, Belgium

Geert Van Hootegem
Center for Sociological Research
Katholieke Universiteit Leuven
Leuven, Belgium

Marcel Veenswijk
Faculty of Social Sciences
VU University Amsterdam
Amsterdam, The Netherlands

Lisa Warnecke
GeoManagement Associates
Syracuse, New York

Section 1

Institutions and Organizations

1

When Is Providing Spatial Information a Public Task? A Search for Criteria

Katleen Janssen, Joep Crompvoets, and Jos Dumortier

CONTENTS

1.1 Introduction

The public sector collects and uses spatial data for preparing, implementing, and evaluating policy and for providing efficient services to citizens (e.g., in the domains of land use, spatial planning, environmental policy, transport, agriculture, public utilities, national defense, and emergency services). However, public sector spatial data are also of great value outside the public sector. Citizens and companies need reliable spatial data for making their decisions on buying a new house or determining the location of a new company branch. In addition, the information industry uses public sector spatial data to create information products or services such as navigation systems, weather reports, or real estate services.

In determining whether it should make its spatial data available to others, the public sector is confronted with different needs and interests. First, there is a growing trend toward openness and transparency of the functioning of public bodies, which includes access to public sector data. One of the main examples of this trend is the adoption of the Aarhus Convention on Access to Information, Public Participation in Decision-Making and Access to Justice in Environmental Matters (United Nations Economic Commission for Europe 1998). A second trend is the growing requirement of efficiency and cooperation within the public sector. Data should be collected only once in the appropriate place and reused by other public bodies whenever possible. Many global, European, and national initiatives are currently being taken to improve this exchange of spatial data, such as the United Nations Spatial Data Infrastructure (UNSDI, United Nations Geographic Information Working Group 2007) and INSPIRE (Infrastructure for Spatial Information in Europe; European Parliament and Council 2007).

Third, the potential economic value of public sector spatial data makes them an attractive resource for the information industry. These data enjoy the assumption of reliability and longevity, and they often have a strong reputation of high quality (Office of Fair Trading 2006).

Finally, many public bodies, particularly in the European Union, have realized that their spatial data are a resource with considerable economic value and they use these data to gain some extra revenue and complement their budget received from central government. Some public bodies are even required to obtain such revenues because they are no longer (completely) funded by the government (Janssen and Dumortier 2007). They can do this in two ways: On the one hand, a public body can charge or impose conditions for the spatial data it provides to the users, including companies, citizens, and public bodies; on the other hand, it may create its own information services or products based on spatial data and sell these added-value products on the market. Examples of such services are dedicated weather services for leisure activities, tourist guides, and navigation services. The latter activity could cause tension between the public and the private sector because concerns may arise that the often dominant position of the public bodies may distort the market (Volman 2004; Weiss 2004).

The different interests in public sector spatial data mentioned before have led to many discussions on the role of public bodies in providing spatial data or offering information services. While the global discussion on the obligation of the public sector to provide spatial data mostly revolves around the need for spatial data for the prevention of and reaction to cross-border environmental and security issues, and the responsibility of governments to provide and use spatial data to ensure the well-being and the welfare of the public (Group on Earth Observations 2005; European Parliament and Council 2007), the debate in the European Union also involves another issue. In the context of the European Union, the debate about the obligations for the public sector to make its data available also centers around the scope of the *public task* of the public sector with regard to providing spatial information services in relationship with the activities of the information industry (Commission of the European Communities 2009).

While the private sector acknowledges the need for high-quality spatial data integrated in accessible services for citizens, it feels that it also has a role to play in the delivery of such services. Hence, the private sector asks the public sector to stay within the limits of its public remit. However, this entails that the limits of this remit—or the public task, as it is referred to in the European Union—should be determined. What data or information should the public bodies make available, and what should they leave to the private sector?

In this paper, we examine whether a definition of the public task or the role of the government is possible. First, we look at why such a definition is called for and needed, against the background of the European Union's legal framework on the availability of public sector spatial data. Next, we try to determine if the scope of the public task can be defined. For this, we examine existing policy documents and literature in this field. This study shows that the public task cannot be clearly defined. In the last part of the paper, some suggestions for resolving the issue of the public task are offered.

1.2 Defining the Role of Government: Why Does It Matter?

The demarcation of the scope of the public task has regularly occurred on the agenda of public bodies and the information industry over the last few years, particularly with regard to the provision of information services by the public sector. The information age has led to a call for rethinking the role of government, particularly with regard to the provision of online information (Stiglitz, Orzag, and Orzag 2000). With the evolution of the information society, the continuously growing amount of information that is available to individuals has not always made their lives easier. Rather, it has become difficult for most citizens to filter the excessive, useless, or faulty information from the information that has the potential to reduce uncertainty and to facilitate decision making (Beers 1996).

In addition, the growing reliance on information has created a digital divide that separates the *information haves* from the *information have-nots* (United Nations Development Program 1999). The government has an important role to play in this evolution by providing information to increase democratic participation and social inclusion, but it also has to stimulate economic growth and allow the private sector to develop information services. To find a balance between the different actors in the information society, a vision on the public task of the government and the division of roles in the information market is vital (Kabel et al. 2001).

In relation with this, a number of other questions about the information market also come into play; the distinction between the public task and the other activities of public bodies is also important with regard to questions of fair competition and a level playing field (Longhorn and Blakemore 2004; Volman 2004). A large part of the data held by public bodies was created in the framework of their public task and cannot be obtained elsewhere. Insofar as a public body offers these data on the market or creates information services or products with them, it is in principle a monopolist. Hence, its behavior can have a considerable impact on the market (Weiss 2004). The possible distortions caused by such behavior may have as a result that the diversity provided by commercial and noncommercial information providers will diminish (Gellman 1996). Therefore, the definition of the public task is also vital for safeguarding the offer of a variety of information services that can respond to the social and economic needs of the society.

1.3 The Importance of Defining the Public Task in the European Union

As was mentioned before, the debate on the public task has become intense in the European Union due to the adoption of legislation that specifically relies on a vision of the tasks of the government to provide spatial information (i.e., Directive 2003/98/EC on the reuse of public sector information—hereafter referred to as "PSI directive"—and Directive 2007/2/EC establishing an infrastructure for spatial information in the European Community—hereafter referred to as "INSPIRE directive"). One of the main criteria for the distinction between these two directives is precisely the concept of the public task.

First, the INSPIRE directive provides the legal basis for setting up a European Community spatial data infrastructure and obliges the member states and the public bodies to share their spatial data with other public bodies for the performance of their "public tasks that may have an impact on the environment" (article 17.1 of the INSPIRE directive). Hence, under this directive, the public bodies have to make their data available, under certain conditions, to other public bodies for the development of their environmental policies regarding, for instance, pollution, climate change, sustainability, and protected sites. Second, the PSI directive addresses the economic value of public sector data and intends to harmonize the conditions for reuse of data held by the public sector. The directive lays down a minimal set of rules for the availability of public sector data for any commercial or noncommercial use outside the public task.

On the one hand, the directive facilitates the use of data by the private sector by imposing conditions on the format of delivery of the data and the charges for the data. On the other hand, it tries to ensure a level playing field on the market by imposing a number of conditions on the public bodies when they create their own information products or services based on their data. The broad concept of reuse as any use for commercial or noncommercial use outside the public task entails that public bodies may also be reusing their own spatial data or spatial data from other public bodies when they are using these to create spatial information services outside their public task.

Hence, the distinction between reuse in the PSI directive and sharing in the INSPIRE directive is based on the concept of the public task. If public bodies are providing information services as part of their "public task with an impact on the environment," the rules of the INSPIRE directive have to be followed. If they are providing these information services for a purpose "other than the initial purpose within the public task" (article 2.4 of the PSI directive), in competition with the private sector on the market, then they have to follow the rules of the PSI directive. This entails that, if the public body uses its own spatial data as a basis for an information service on the market, it has to make these spatial data available to its competitors under the same conditions and charges. However, the concept of public task is not defined in either of the directives, so it remains unclear where the distinction between the PSI directive and the INSPIRE directive lies.

The question of the public task to provide information services is not just a matter of the relationship between the two directives at the European level. It has also come up at the national level. For instance, the Dutch minister of internal affairs already asked in 1997 to what extent providing information services was a market activity (Ministry of the Interior 1997). The activities of many public bodies have been questioned since then—for example, *Rijkswaterstaat* (Van Loenen et al. 2006, 2007; Van Eechoud and Van der Wal 2008), Royal Dutch Meteorological Service (KNMI) (KPMG 2007), Ordnance Survey (Office of Fair Trading 2006; Communities and Local Government Committee 2008; Saxby 2008), the Swedish National Land Survey (*Lantmateriet*) (NautaDutilh 2004; Statskontoret 2005), and the French Meteorological Office (Bruguiere 2002; Teresi 2005).

In summary, the question of the role of the government or its public task to provide spatial information services to the public is important on a conceptual level—to have a vision on what can and should be expected of government—as well as on a more practical level—to determine the applicable EU legislation for a particular information service provided by the public sector and to decide whether a particular national service is performed in competition with the private sector or not. In the following section, we try to define criteria that indicate whether the government has a public task to provide an information service, mainly from the perspective of the separation of the roles of the government and the private sector on the spatial information market and of the application of the PSI and the INSPIRE directives.

1.4 Defining the Public Task: Finding Criteria

The importance of a clear view on the scope of the public task, particularly from the perspective of knowing the applicable EU legislation, requires that this public task should be defined. In this section, it is examined whether such a definition is possible. First, some attention is given to the existing obligations in legislation and policy documents for public bodies to provide information services to the public. The focus is on the provision of services that go beyond merely making the data available, as this is the domain where the main discussion on the public task takes place. After this short overview, we examine the main criteria for determining a public task to provide information services that are offered in the literature.

1.4.1 Existing Obligations

The existing obligations for the public sector to provide information services may already give the first indication of what international organizations and the European Union consider to be part of the public task of the public sector to provide information services. Over the last few decades, public bodies have been subject to increasing obligations, both on a national and an international level, not only to make their data available to the general public, but also to inform citizens about matters that may concern them. Hence, their public task of making data available and providing information services is growing—for example, in the Aarhus Convention (United Nations Economic Commission for Europe 1998), the Council of Europe Treaty on Access to Official Documents (Council of Europe 2008), and the Directive on Access to Environmental Information (European Parliament and Council 2003).

However, in general, the European and international legislator and policy maker do not have an explicit opinion on how far the public bodies need to go in informing the public. The main obligations that can be found are requirements for the public bodies to make information "effectively accessible" (Aarhus Convention), to promote "adequate dissemination" (OECD Recommendation on Environmental Information, Organization for Economic Cooperation and Development 1998), or to provide "comprehensible" information (Directive on Access to Environmental Information), which are all obligations that are dependent not only on the efforts of the public body, but also on the citizens who receive the information.

Only a few concrete obligations regarding the provision of information on the functioning of the public bodies can be traced—for example, on how to exercise the right of access and on the creation of reports on the state of the environment (Aarhus Convention, Access Directive). On the national level, some more detailed obligations can be found, such as the Dutch law on the public task of the Royal Meteorological Service. However, the question can be asked as to how valid these descriptions are. In the next section we will discuss the possible criteria for defining the public task.

1.4.2 Criteria

The obligations for public bodies to provide information services in current legislation are a good starting point in determining the scope of the public task of the public sector to provide information services. However, these obligations are not very clear and they are only a snapshot in time of the legislators' and policy makers' priorities. Therefore, it is important to determine on which criteria these priorities are based, in order to have a more encompassing view on the scope of the public task. In the literature, several criteria can be found for accepting a public information task.

1.4.2.1 Legal Basis

The first criterion is directly linked to the existing legislation and policy documents mentioned earlier. The most straightforward criterion to determine whether an activity of a public body forms part of its public task or not is the presence of a legal basis (Hengstschlager 1995; Nouel 1996; Di Fabio 1999; Wetenschappelijke Raad voor het Regeringsbeleid 2000). This entails that a public body's public task can be defined as the obligations that are assigned to this public body by the law. This legal basis does not necessarily have to be a law that is voted by parliament; it can also be an administrative decision.

From this perspective, the INSPIRE directive would apply to the provision of information services that are described in the law, decree, or executive decision that determines the role and tasks of a particular public body. If such an information service cannot be deduced from the obligations of the public body in the official text, the service would fall under the PSI directive and possibly be in competition with the private sector.

There are several problems with this criterion. First, while the incorporation of the public task in a law should provide for legal certainty and democratic legitimacy, the extensive and time-consuming procedure for changing such a law makes this public task rigid and unable to be adapted to changing circumstances. In order to avoid a quickly outdated description of the public task, it should be kept at a very high level, only setting out the main elements of the public task. However, such a general description does not specify in detail the role of the public sector. For instance, describing the task of a national mapping agency as "providing topographic mapping for the national territory" does not help in determining the way the mapping should be done, the way the data should be presented, or which services can be provided based on the data.

This vague character of the public task could be decreased if it were written as an administrative decision rather than a law because this format would leave more room for flexibility. However, this approach also has a number of drawbacks. First, the easy adaptability of such administrative decisions makes them vulnerable to changes based purely on political or personal convictions. One could question whether this risk of unpredictable changes would discourage the private sector from offering comparable information services in the market. In addition, concern could arise that the public body providing an information service could determine its own public task, thereby effectively preventing competition on the market (Communities and Local Government Committee 2008).

This relates to the next difficulty of using a legal basis as the criterion for determining the public task: It leads to a circular argument. An activity of a public body is considered a public task, so it is laid down in the law as such (KPMG 2007); these legally recognized tasks are in turn considered public tasks because they are laid down in the law. This still does not explain why the legislative or administrative body considered a particular task to be a public task that should be written into law in the first place.

1.4.2.2 Link with the Core Responsibility of the Public Body

Some literature does not see the provision of information services as a separate public task, but relates it to the other activities of the public body. The authors claiming this believe that an information service is part of the public task if it is an extension of the public body's public service activities (Nouel 1996), if it has a close link with the public sector missions (Burkert 1995), or if it is related to the public body's core responsibility (Office of Public Sector Information 2005). However, this approach only moves the problem to another realm; it does not solve it. The core responsibilities or public service activities still have to be identified before the link with the information service comes into play. These core responsibilities will in turn be determined by the same criteria that are discussed in this section on possible criteria for determining the public task.

1.4.2.3 Market Failure

The third criterion for the public bodies to create spatial information services is often found in the concept of market failure (i.e., the market is not providing spatial information services that the policy maker feels should be provided). There may be several reasons why these spatial information services are not offered on the market. For instance, the private sector may refuse to provide an information service because market demand is insufficient to be worthwhile for the private sector (Van Damme 1999; Huisman and Van De Lei 2003). Market failure may also occur if the desired spatial information service is offered by the private sector, but at a higher cost or lower productivity than the government would want, or if it endangers pluriformity (Heijne 1995; Stiglitz et al. 2000; Dijstelbloem and de Beer 2003).

Market failure has repeatedly been considered a reason for government intervention in the provision of information services (Kabel et al. 2001;

Sears 2001; Bruguiere 2002). With regard to the EU legal framework, this would mean that a public body providing a spatial information service that the market could or would not provide can obtain its data resources under the INSPIRE directive. Public bodies creating services that the market can offer would be subject to the PSI directive.

However, this criterion can also give rise to some remarks. First, the response of the public sector to market failure may take different forms: imposing regulations to ensure the adequate provision of the information service by the private sector, awarding subsidies to the private sector, offering the service on the market, or offering the service as a public task. Hence, it does not necessarily imply that the information service will be offered as a public task. Moreover, the fact that the market does not offer a particular information service does not always mean that the public body should get involved. Market inactivity does not always equal market failure (Zwenne 1997). The public body or the government still has to decide that the information service is really needed to meet a particular societal need or interest.

1.4.2.4 Public Goods

Related to the theory of market failure is the concept of public goods. Such goods are in principle only produced or offered by the public sector because they would otherwise be undersupplied, as it is not profitable for the private sector to provide them (Stiglitz 1999). Public goods are characterized by two attributes: They are nonexcludable and nonrivalrous. The former entails that it is difficult, if not impossible, to exclude an individual from enjoying the good, while the latter means that the consumption of the good by one individual does not deter another person from consuming it. Classic examples of public goods include national defense, street lighting, and environmental protection. Information in general and public sector spatial data in particular are also often considered as a public good (Onsrud 1998; Nilsen 2007).

Applying the public good perspective to the relationship between the PSI directive and the INSPIRE directive is difficult. If spatial information is a public good, one possible interpretation of the directives could lead to the adoption of a very wide public task under the INSPIRE directive without consideration of the type of service or the target audience of the service. This would risk making the PSI directive obsolete with regard to spatial information because such information services would never be considered a market service outside the public task.

Some nuances should also be brought to the concept of spatial information as a public good. While information may be a public good, the way it is disseminated may create the possibility to exclude people from consuming it or make it contentious (e.g., with passwords, encryption, intellectual property rights, or digital rights management). Moreover, as public sector spatial data are increasingly being produced by the private sector, their public good character is also relative. Therefore, some authors suggest considering public sector information or spatial data as a merit good: a good that is subsidized by the public sector because its existence or consumption is highly valued by the community (Love 1995; Kabel et al. 2001). Government has a role in the provision of such goods, either by providing the goods or by ensuring that the private sector does so (Batley 1996). Hence, the same problem arises as with regard to market failure. It still has to be decided which goods are highly valued (i.e., which goods are in the public interest).

1.4.2.5 Public Interest

The notion of public interest leads to the final criterion generally used for determining the scope of the public task. Public bodies should provide spatial information services when they are needed for the pursuit of the public interest. The public interest can take many forms, including ensuring fair and democratic policy (Baten and Van Der Starre 1996); protecting democratic rights (Dommering et al. 2002); guaranteeing the reliability, neutrality, or pluriformity of information (Lamouline 1995; Nouel 1996; Bovens 2002; Dijstelbloem and De Beer 2003); and safeguarding environmental concerns, public health, and social capital (Longworth 2000).

With regard to information services, public interest could be ensured by or translated in accessibility. This means that citizens can use information services with a relatively small effort and with minimal barriers (Janssen, Steyaert, and van Gompel 2003) in order to pursue their needs and interests in an adequately informed way. Hence, if the public is entitled to accessible information provided by the government, the INSPIRE directive applies to the data resources needed for the information service. If this is not the case, the PSI directive applies.

However, the concept of accessibility immediately shows the problem with the concept of the public interest. Accessibility holds several aspects. The information has to be physically accessible, financially affordable, and intellectually comprehensible (Beers 1996; Bouwman et al. 1996). For all these aspects, the subjective character of accessibility should be recognized. It depends on the practical or technological means of access available to the target group, the level of education and skills of the receiver, or the financial resources available to pay the possible charges for the information service. Moreover, in increasing the accessibility to its data, the public body should be careful that by placing data into a context and hence making it become information to users, it is not tempted to manipulate the resulting information and influence the public (Baten and Van der Starre 1996; Wiese Schartum 1998). Next, it should also pay attention to its target groups. If the activity supported by public information goes beyond the public interest needs of a target group, it may no longer be performing a public task, but rather acting as a market player in competition with the private sector. Yet the needs of such target groups may change over time or shift in location.

The remarks made about accessibility are also applicable to the general concept of the public interest. While it is essentially the basis for defining the scope of the public task, the public interest is very much dependent on evolutions in technology, socioeconomic circumstances, personal opinions, and political tendencies. Consequently, the public task is also an evolving and political concept (Naschold and Von Otter 1996; Tanzi 2000; Reichard 2006). It depends on technological development, changes in user expectations, economic developments, and cultural legacies, among other factors (Rockman 1997; Wetenschappelijke Raad voor het Regeringsbeleid 2000; Vermeulen 2003).

Moreover, the views of the public task may differ between countries: Information on flooding risks or the rise of water levels due to climate change will be more relevant for coastal countries than for countries lying inland (Prosser 2005; Steyger 2007). The political character of the public task increases this evolving character of the public task. For instance, the view on the core tasks of the public sector was different in the classical minimalist or liberal concept of government at the end of the nineteenth century from the notion of the welfare state of the second half of the twentieth century (Shamsul Haque 2001; Brown 2003). More recently, the privatization movement of the 1980s and 1990s has been contradicted by the growing intervention of the state as a regulatory influence in a state-regulated welfare market (Wetenschappelijke Raad voor het Regeringsbeleid 2000).

Hence, the political and evolving character of the public task makes it difficult to define it in a sufficiently clear manner for it to be used as a straightforward boundary between the activities of the public bodies and the private sector. Thus, the demand of the information industry for a definition of the public task of the public bodies cannot be answered in a way that completely provides legal certainty and economic security in the long term. Other solutions need to be considered to ensure that the private sector is not discouraged from creating information services and that the European information market can be developed.

1.5 No Definition of the Public Task—What Now?

1.5.1 Empirical Evidence

First, more empirical evidence is needed with regard to the role of government in providing information services to the public. Is the lack of clarity about this role really causing a problem for the availability of spatial information services to citizens or for the growth of the information industry, or does the problem only exist in the minds of a few disgruntled stakeholders? In the European Union, the main concern is the relationship between the public and the private sectors. Assessing whether this relationship is problematic requires determining with more certainty if the concern of the private sector about public bodies performing market activities is valid and if the problem goes deeper than a few isolated cases of discussion.

Since 1996, reports on the European market for information services and products have shown that the division of roles between the public and the private sectors has been problematic (PIRA International 2001; MICUS Management Consulting 2008). During its review of the PSI directive, the European Commission found that stakeholders frequently cited as problematic the limits of the public task when public bodies commercially compete with private firms and unfair competition practices by public sector bodies (Commission of the European Communities 2009). The commission gave the example of the German State Survey Authorities, who consider the production of maps for leisure activities to be a public task and sell them at very low prices, while the public task character of this activity is questioned by the publishing sector.

While there are many questions about the public task, more empirical material should be collected to establish whether the role of the public sector in the information market needs to be addressed in a general manner. The first indication could be the number of complaints and court cases that involve this issue. Complaints and investigations have, for instance, been introduced against the information service activities of the UK Environment Agency (this complaint was found unjustified), the Belgian Meteorological Office (the complaint has not led to a decision by the competition authority), the French Meteorological Office (the complaint was found unjustified), the French Hydrographical Office (the complaint was deemed justified), and the Swedish Land Survey (the competition authority felt that the activities of the land survey should be separated) (Janssen 2010).

However, one should recognize that the actual complaints could be only an indication of a more widespread problem because many small and medium enterprises will not have the means to start a procedure. Finally, in gathering of the empirical evidence, it should be kept in mind that many private sector service providers are dependent on the public bodies for the delivery of their information, so good cooperation between the two sectors should remain the focal point.

1.5.2 Reduction of the Uncertainty of the Public Task

If there is sufficient empirical evidence that the role of the government or the concept of the public task to provide spatial information services is truly problematic, different options should be considered to address this problematic character and the uncertainty it causes. While it has to be accepted that the public task will always remain an evolving and political concept, discussions may be easier on a smaller scale, with regard to particular types of information services or products. It should be considered whether it is possible to determine which services should definitely be included under the public task and which services can easily be ruled as outside of the public task. The unavoidable remaining gray zone can then be dealt with.

This gray zone will need careful consideration, and its definition cannot be left to the government itself. It needs to be subjected to the democratic process, so that the legitimacy of any decision on the public task is guaranteed and, if needed, the decisions can be challenged by the public through its political representatives. In addition, more direct participation of the public should be organized to ensure that the government's idea of the public task reflects the stakeholders' opinion. Such a dialogue could be organized by sector or by type of activity, but it should in any case involve stakeholders with a diversity of backgrounds and opinions, including the public bodies, the private sector, citizens, and nongovernmental organizations (De Terwangne 2001).

The representation of the stakeholders could also take a more permanent form, in an advisory or regulatory body that can provide guidance to the government on its role and its relationship with citizens on the one hand and the private sector on the other hand. This participatory process should ensure a balanced view on the information society and the information market, taking into account the business models of the private sector and the need for the public bodies to ensure public interest, and leaving room for innovation and adaptation to the changing needs of society.

As the scope of the public task to provide information services is becoming an issue on the European and even the global level, the discussion on the role of the government cannot be limited to the national level. Global agreement should be attempted, particularly with regard to information services that are recognized by everyone as part of the public task. However, considering the close link of the views on the role of government to national politics and traditions, such global agreement will have to grow incrementally.

In the particular situation of the European Union—where the public task is an explicit basis for the field of application of the PSI directive and the INSPIRE directive—the discussion on the scope of the public task and the role of government would also benefit from a broader discussion with all stakeholders and a search for a *European Union public task* concept. However, the European Commission has already indicated that it does not intend to provide any guidance for such a discussion and that it remains up to the member states to define their public task (Commission of the European Communities 2009). Yet the inclusion of the public task as a criterion for the application of concrete legislation makes the debate more tangible and increases the chances that this criterion is actually applied because both the public bodies and the private sector are demanding clarification on the issue. Hence, in the European Union, an advisory group or body on the public task may have an opportunity to produce useful guidelines or directions for the activities of the government in the information sector.

1.6 Conclusion

In this chapter, the main criteria offered in the literature for determining the public task were examined against the debatable nature of provision of spatial information under the European Union PSI and INSPIRE directives. It was found that the suggested criteria could all be traced back to the public interest. An activity of a public body could be considered to be a public task when it is answering a need in the public interest. However, this public interest depends on many different factors in society, the economy, and politics, which entails that the public task also has an evolving and political character. This makes it difficult to formulate an operational and simple definition of the public task.

However, the uncertain character of the public task could be reduced by using the democratic process and stakeholders' involvement to attempt to agree on information services that are undoubtedly part of the public task and services that should definitely be left to the private sector. The involvement of all the stakeholders should guarantee a sufficiently balanced view on the public task to be acceptable by both the private and the public sectors. However, both sectors should still understand that this can only limit the uncertainty, but never remove it completely.

Inevitably, a gray zone will remain that requires more in-depth discussion between the stakeholders and that may only be cleared up by case-by-case decisions. In order to further the discussion on the role of the government in the provision of spatial information services, this gray zone will have to be addressed in cooperation with all the stakeholders, while also taking into account the current political, economic, social, and technological circumstances that may influence the development of theory and practice with regard to this topic.

References

Baten, I., and G. Van Der Starre. 1996. *Elektronische toegankelijkheid van overheidsinformatie.* The Hague: Rathenau Instituut.

Batley, R. 1996. Public–private relationships and performance in service provision. *Urban Studies* 33:723–751.

Beers, A. L. 1996. Openbaarheid van overheidsinformatie. In *Elektronische toegankelijkheid van overheidsinformatie,* ed. I. Baten and G. Van Der Starre, 51–78. The Hague: Rathenau Instituut.

Bouwman, H., J. van Cuilenburg, P. Neijens, and J. Nouwens. 1996. Leuker kunnen we het niet maken, wel makkelijker. Over de toegankelijkheid van overheidsinformatie. In *Elektronische toegankelijkheid van overheidsinformatie,* ed. I. Baten and G. Van Der Starre, 79–88.The Hague: Rathenau Instituut.

Bovens, M. 2002. Information rights: Citizenship in the information society. *Journal of Political Philosophy* 10:317–341.

Brown, G. 2003. State and market: Towards a public interest test. *Political Quarterly* 74:266–284.

Bruguière, J. M. 2002. *Les données publiques et le droit.* Paris: Editions Litec.

Burkert, H. 1995. Conclusions. In *Proceedings of the Workshops on Commercial and Citizen's Access to Government Information,* ed. Commission of the European Communities, 80–82. Luxembourg: Commission of the European Communities.

Commission of the European Communities. 2009. Staff working document. Accompanying document to the communication from the Commission to the European Parliament, the Council, the European Economic and Social Committee and the Committee of the Regions on the reuse of public sector information. http://ec.europa.eu/information_society/policy/psi/docs/pdfs/directive/com09_212/staff_working_document.pdf (accessed November 16, 2009).

Communities and Local Government Committee. 2008. Ordnance survey, fifth report of session 2007–08. http://www.publications.parliament.uk/pa/cm200708/cmselect/cmcomloc/268/26802.htm (accessed November 16, 2009).

Council of Europe. 2008. Convention on access to official documents. https://wcd.coe.int/ViewDoc.jsp?id=1377737&Site=CM (accessed November 16, 2009).

De Terwangne, C. 2001. *Société de l'information et mission publique d'information.* Namur: Facultés Universitaires de Notre Dame de Paix.

Di Fabio, U. 1999. Privatisierung und Staatsvorbehalt. *Juristen Zeitung* 12:585–592.

Dijstelbloem, H., and P. De Beer. 2003. Een kwestie van selectie: de overheid en de informatievoorziening via internet. *Beleid en Maatschappij* 30:242–250.

Dommering, E. J., P. B. Hugonholtz, and J. J. C. Kabel. 2002. De overheid en het publiek domein van informatie voor wetenschappelijk onderzoek. In *De publieke dimensie van kennis,* ed. H. Dijstelbloem and C. Schuyt. The Hague: WRR.

European Parliament and Council. 2003. Directive 2003/4/EC on public access to environmental information and repealing council directive 90/313/EEC. *Official Journal of the European Union* 41:26–32.

European Parliament and Council. 2007. Directive 2007/2/EC establishing an infrastructure for spatial information in the European Community. *Official Journal of the European Union* 108:1–14.

Gellman, R. 1996. The American model of access to and dissemination of public information. In *Proceedings of the European Commission Conference: Access to Public Information. A Key to Commercial Growth and Electronic Democracy,* ed. Commission of the European Communities, 1–11. Luxembourg: Commission of the European Communities.

Group on Earth Observations. 2005. The global earth observation system of systems (GEOSS) 10-year implementation plan, www.earthobservations.org/documents/10-Year%20Implementation%20Plan.pdf (accessed February 22, 2010).

Heijne, G. 1995. Perspectieven op informatiepolitiek. Een verslag gebaseerd op vier workshops. In *Een kwestie van toegang. Bijdragen aan het debat over het publieke domein van de informatievoorziening,* ed. I. Baten and J. Ubacht, 45–61. The Hague: Rathenau Instituut.

Hengstschlager, J. 1995. Privatisierung von Verwaltungsaufgaben. In *Veröffentlichungen der Vereinigung Deutscher Staatsrechtlehrer 54,* 165–203. Berlin: De Gruyter.

Huisman, K., and J. Van De Lei. 2003. Wet FIDO en publieke taak onder Wet Dualisering. *Tijdschrift B&G*:34–35.

Janssen, K. 2010. *The availability of spatial and environmental data in the EU at the crossroads between public and economic interests*. Dordrecht: Kluwer.

Janssen, K., and Dumortier, J. 2007. Legal framework for a European Union spatial data infrastructure: Uncrossing the wires. In *Research and theory in advancing spatial data infrastructure concepts*, ed. H. J. Onsrud, 231–244. Redlands, CA: ESRI Press.

Janssen, K., J. Steyaert, and R. van Gompel. 2003. *Transparantie van overheidsinformatie in Vlaanderen. Beschouwingen en aanbevelingen*. Bruges: Die Keure.

Kabel, J. J. C. et al. 2001. *Kennisinstellingen en informatiebeleid. Lusten en lasten van de publieke taak*. Amsterdam: Universiteit van Amsterdam.

KPMG. 2007. Ministerie van Waterstaat. Evaluatie van de wet op het KNMI. http://www.verkeerenwaterstaat.nl/Images/2008651%20bijlage_tcm195-231492.pdf (accessed November 16, 2009).

Lamouline, C. 1995. Presentation of PUBLAW 3 findings. Legal assessment. In *Proceedings of the Workshops on Commercial and Citizen's Access to Government Information*, ed. Commission of the European Communities, 9–14. Luxembourg: Commission of the European Communities.

Longhorn, R., and M. Blakemore. 2004. Revisiting the valuing and pricing of digital geographic information. http://journals.tdl.org/jodi/article/view/103/102 (accessed November 16, 2009).

Longworth, E. 2000. The role of public authorities in access to information: The broader and more efficient provision of public content. Study prepared for the Third UNESCO International Congress on Ethical, Legal and Societal Challenges of Cyberspace INFOethics. http://unesdoc.unesco.org/images/0012/001210/121051e.pdf (accessed November 16, 2009).

Love, J. 1995. Pricing of government information. *Journal of Government Information* 22:363–387.

MICUS Management Consulting GMBH. 2010. Assessment of the Re-use of Public Sector Information (PSI) in the Geographical Information, Meteorological Information and Legal Information Sectors, http://ec.europa.eu/information_society/policy/psi/docs/pdfs/micus_report_december2008.pdf (accessed 17 May 2009).

Ministry of the Interior. 1997. Naar toegankelijkheid van overheidsinformatie. http://www.minbzk.nl/aspx/download.aspx?file=/contents/pages/302/nota_ntvo_nl.pdf (accessed November 16, 2009).

Naschold, F., and C. Von Otter. 1996. *Public sector transformation. Rethinking markets and hierarchies in government*. Amsterdam: John Benjamins Publishing Company.

NautaDutilh. 2004. *Dealing with dominance: The experience of national competition authorities*. Amsterdam: Kluwer Law International.

Nilsen, K. 2007. Economic theory as it applies to statistics Canada: A review of the literature. http://www.chass.utoronto.ca/datalib/misc/Nilsen%20Economics%20Paper%202007%20final%20version.pdf (accessed February 25, 2009).

Nouel, B. 1996. Commercialisation des données publiques et concurrence. In *Proceedings of the European Commission Conference: Access to Public Information. A Key to Commercial Growth and Electronic Democracy*, ed. Commission of the European Communities, 1–9. Luxembourg: Commission of the European Communities.

Office of Fair Trading. 2006. The commercial use of public information (CUPI). http://www.oft.gov.uk/shared_oft/reports/consumer_protection/oft861.pdf (accessed November 16, 2009).

Office of Public Sector Information. 2005. The Re-use of Public Sector Information: A Guide to the Regulations and Best Practice, www.nationalarchives.gov.uk/.../guide-to-psi-regulations-and-best-practice.doc (accessed 22 July 2010).

Onsrud, H. 1998. Tragedy of the information commons. In *Policy issues in modern cartography*, ed. D. R. F. Taylor, 141–158. Oxford, England: Elsevier Science Ltd.

Organization for Economic Cooperation and Development. 1998. Recommendation of the Council on Environmental Information. http://webdomino1.oecd.org/horizontal/oecdacts.nsf/linkto/C(98)67 (accessed on February 22, 2010).

PIRA International. 2001. Commercial exploitation of Europe's public sector information—Final report. http://ec.europa.eu/information_society/policy/psi/docs/pdfs/pira_study/commercial_final_report.pdf (accessed November 16, 2009).

Prosser, T. 2005. *The limits of competition law. Markets and public services.* Oxford, England: Oxford University Press.

Reichard, C. 2006. New institutional arrangements of public service delivery. In *The governance of services of general interest between state, market and society*, ed. C. Reichard et al., 35–47. Berlin: Wissenschaftlicher Verlag Berlin.

Rockman, B. 1997. Honey, I shrank the state. In *Modern systems of government. Exploring the role of bureaucrats and politicians*, ed. A. Farazmand, 275–295. London: SAGE Publications Inc.

Saxby, S. 2008. Public sector information and reuse policy—Where is the UK now? Second International Conference on Business, Law and Technology. http://eprints.soton.ac.uk/52523/01/PSI_Re-Use_Paper_28_April_2008_pdf.pdf (accessed November 16, 2009).

Sears, G. 2001. Geospatial data policy study. Executive summary. http://www.geoconnections.org/programsCommittees/proCom_policy/keyDocs/KPMG/KPMG_E.pdf (accessed November 16, 2009).

Shamsul Haque, M. 2001. The diminishing publicness of public service under the current mode of governance. *Public Administration Review* 61:65–82.

Statskontoret. 2005. Competition at the public/private interface. http://www.statskontoret.se/upload/Publikationer/2005/200519A.pdf (accessed November 16, 2009).

Steyger, E. 2007. Deel II: De publieke taak: Diensten van algemeen (economisch) belang en de gevolgen van de richtlijn. *Sociaal-Economische Wetgeving* 10:379–398.

Stiglitz, J. 1999. Knowledge as a global public good. In *Global public goods*, ed. I. Kaul et al., 308–325. Oxford, England: Oxford University Press.

Stiglitz, J., P. R. Orzag, and J. M. Orzag. 2000. The role of government in a digital age. http://unpan1.un.org/intradoc/groups/public/documents/apcity/unpan002055.pdf (accessed November 16, 2009).

Tanzi, V. 2000. IMF working paper. The role of the state and the quality of the public sector. http://www.imf.org/external/pubs/ft/wp/2000/wp0036.pdf (accessed November 16, 2009).

Teresi, L. 2005. Données publiques et concurrence. *Concurrences* 4:143–156.

United Nations Development Program. 1999. Human development report 1999. http://hdr.undp.org/en/media/HDR_1999_EN.pdf (accessed on February 22, 2010).

United Nations Economic Commission for Europe. 1998. Convention on Access to Information, Public Participation in Decision-making and Access to Justice in Environmental Matters. http://www.unece.org/env/pp/documents/cep43e.pdf (accessed November 16, 2009).

United Nations Geographic Information Working Group. 2007. UNSDI compendium: A UNSDI vision, Implementation Strategy and Reference Architecture. www.ungiwg.org/docs/unsdi/UNSDI_Compendium_13_02_2007.pdf (accessed February 22, 2010).

Van Damme, E. 1999. Concurrentievervalsing door de overheid? *Openbare Uitgaven* 30:194–203.

Van Eechoud, M., and B. Van der Wal. 2008. Creative commons licensing for public sector information. Opportunities and pitfalls. http://www.ivir.nl/publications/eechoud/CC_PublicSectorInformation_report_v3.pdf (accessed November 16, 2009).

Van Loenen, B., F. Welle Donker, H. Ploeger, and J. Zevenbergen. 2006. Beschikbaar stellen van geo-informatie bij Rijkswaterstaat. Analyse van de (on)mogelijkheden van het op korte termijn vrij beschikbaar stellen van vier geo-data sets. http://www.bastiaanvanloenen.nl/pubs/2006_RWS_AGI_Eindrapport_TUD.pdf (accessed November 16, 2009).

Van Loenen, B., J. Zevenbergen, G. Giff, and J. Crompvoets. 2007. Open toegankelijkheidsbeleid voor geo-informatie vergeleken: het gras leek groener dan het was. http://www.minbzk.nl/aspx/download.aspx?file=/contents/pages/87285/opentoegankelijkheidsbeleidvooroverheidvergeleken.pdf (accessed November 16, 2009).

Vermeulen, B. 2003. De publieke taak: Een veel-zijdig begrip. In *De publieke taak. Staatsrechtconferentie 2002, Vrije Universiteit Amsterdam,* ed. J. Sap et al., 13–30. Amsterdam: Kluwer.

Volman, Y. 2004. Exploitation of public sector information in the context of the eEurope action plan. In *Public sector information in the digital age. Between markets, public management and citizens' rights,* ed. G. Aichholzer and H. Burkert, 93–107. Cheltenham, England: Edward Elgar Publishing Limited.

Weiss, P. N. 2004. Borders in cyberspace: Conflicting public sector information policies and their economic impacts. In *Public sector information in the digital age. Between markets, public management and citizens' rights,* ed. G. Aichholzer and H. Burkert, 93–107. Cheltenham, England: Edward Elgar Publishing Limited.

Wetenschappelijke Raad voor het Regeringsbeleid. 2000. *Het borgen van publiek belang.* The Hague: Sdu Uitgevers.

Wiese Schartum, D. 1998. Access to government-held information: Challenges and possibilities. *Journal of Information, Law and Technology* 1. http://www2.warwick.ac.uk/fac/soc/law/elj/jilt/1998_1/schartum/ (accessed November 16, 2009).

Zwenne, G. 1997. De informatieconsument en de markt voor overheidsinformatie. *Tijdschrift voor Consumentenrecht* 3:223–231.

2

Institutionalization Does Not Occur by Decree: Institutional Obstacles in Implementing a Land Administration System in a Developing Country

Leiser Silva

CONTENTS

ABSTRACT This paper focuses on the implementation of a spatial data infrastructure (SDI) in a developing country. It concentrates on the institutional obstacles of such implementations. Given the heterogeneity of data sources required for SDIs to function, inter-institutional cooperation is a fundamental condition for these infrastructures to operate successfully. It is argued that one of the main obstacles for institutional cooperation is the contradicting meanings assigned to an SDI by users and designers. Those contradicting meanings were identified through an interpretive study conducted in Guatemala. Using hermeneutic analysis those meanings were synthesized into four main themes: (1) the contradicting views of Guatemalans and Europeans regarding the rationality of institutions, (2) the link between SDIs and work tasks, (3) institutional jealousy and (4) the historical resistance to re-examine the institution of land ownership in Guatemala. Each of these themes is examined in the light of critical theory. In so doing, implications for practice and research are discussed. The paper concludes by proposing a guideline based on actor-network theory (ANT) that suggests a set of steps required to achieve institutional cooperation.

KEYWORDS: SDIs, land administration systems, developing countries, Guatemala, power, politics, implementation of information systems, institutions, institutionalization.

2.1 Introduction

Land administration systems supported by spatial data infrastructures (SDIs) are deemed to be fundamental means for establishing a basis of prosperity in developing countries.* They have been linked to the economic development, environment conservation and social stability of nations [32,67,74,76]. Indeed, the information and services provided by land administration systems are essential for implementing land use policies that range from environmental conservation to conflict solving about land ownership. The latter is of a particular relevance for the social stability of post-conflict developing nations such as Nicaragua, Kosovo or El Salvador [74]. Implementing SDIs in such contexts is a challenge due to the complexity of the technologies as well as the instability of their institutions. In Nicaragua, for example, one of the main sources of conflict after the Sandinista revolution has been determining the ownership of properties. The problem is exacerbated given that large amounts of property titles were destroyed and the institutions† in charge of determining and refereeing conflicts lack resources and operate within a weak legal framework.

* Even though SDIs are the technological infrastructures of land administration systems, the term will be used indistinctively, since whether it is referred to the technological or application dimension is a matter of perspective. For the users these are land administration systems and for the designers and engineers they may be just SDIs.

† In this paper the term institution is used to refer to formal public organizations [36].

Although SDIs' implementation in developing countries has been central in some of the literature [32,37,38,69,74,75], little is known about how these systems become institutionalized. In the context of this paper, the term institutionalization is used to refer to the adoption of systems by organizations; an institutionalized system becomes routinized in the practices and procedures of public organizations [65]. As will be discussed further, the institutionalization of a land management system is even more complex given that it implies cooperation among different institutions. Accordingly, the objective of this paper is to identify obstacles that hinder the institutionalization of a land management system. Seen this way, the process of institutionalization implies the exercise of power, alignment and enrolment of different actors, as well as the creation of alliances [63]. It will be argued in a further section of the paper that the institutionalization of a system can be understood as the creation of an actor-network [7,9,64].

This study is intriguing since the Guatemalan government, with the support of the international community, has been struggling to implement an SDI for the last ten years. Such an initiative began when the Guatemalan authorities formally committed to implement a land administration system as part of the peace accords that ended the Guatemalan civil war in 1996. However, almost ten years after having signed the accords, and in spite of multiple implementation efforts, the land administration system is far from complete [18]. The system has not been institutionalized as the decree established by the Guatemalan government has not been observed by the different institutions that oversee land administration in the country [18]. One of the main reasons for the delay has been the difficulty in coordinating the different Guatemalan institutions associated with the land administration system [18,52].

In addressing the research objective, it is argued that the institutionalization of a system can be better understood from the point of view of power [12]. The assumption is that power operates by actors pursuing their particular interests by reflecting on and interpreting their own situations [9,22]. Therefore, this paper highlights the different and even contradicting meanings that actors assign to a land administration system. The contradicting meanings are made sense of through a hermeneutic interpretation of the data [68] and by looking at it through the lens of critical theory [20,47]. Through its emphasis on revealing the social and political aspects of technology, critical theory allowed us to bring to the discussion the history and context in which a land administration system is interpreted.

The paper is structured in four main sections. This introduction is followed by the concepts of SDIs, land administration and critical theory. Next, the research approach is presented, which, as alluded to above, consisted in conducting an interpretive study. Subsequently, the results of the interpretations are introduced. These consist of four main themes discussed critically and in the light of their theoretical and practical implications. The themes are followed by a proposed model for implementing a land administration system based on sociology of translation and actor network theory (ANT; see Latour [41], Callon [10], Law [43]). This model outlines political maneuvers that may be required to reach institutionalization. As it will be argued, these maneuvers center more on agreements and negotiations rather than on conflict. The paper concludes with a reflection on the implications and limitations of the research, as well as suggesting some areas that may be worth pursuing in future research.

2.2 Theoretical Background: SDIs and Critical Theory

2.2.1 Institutionalization of SDIs and Inter-organizational Cooperation

The notion of SDIs originated in Europe (the UK and The Netherlands) and North America in the late 1970s [25]. Groot observes that SDIs emerged as a response to recent technological advances, such as satellite and digital networks, as well as to a growing demand from users for services involving maps and spatial data. Although some authors like Phillips et al. [57] consider SDIs to be mainly computer equipment, specializing in the storage, process and transmission of spatial and map data, other authors include the policies and processes that feed data to those infrastructures [37]. Thus, Karikari et al. [37] are emphatic in stating that an SDI cannot be implemented without inter-institutional cooperation (see also [70,71,74]).

The inter-organizational nature of SDIs is manifested by the different data sources that provide its input. Typically, spatial data and maps are produced by different organizations within the scope of specific projects. Therefore, the SDI would serve as a repository of such data so it can be used by other organizations or projects without the need to duplicate efforts of data production to achieve the same end results. The synergistic effect and cost-savings of SDIs are then apparent considering the expense involved to produce spatial data [57]. Nonetheless, the sharing of data among different organizations would require these to develop and adopt standards which cannot be attained without intense coordination and negotiations among the different institutions involved [30]. Once this has been achieved, a land administration system may provide services to different organizations such as the management of land titles for property registries, maps of protected areas to environmental related organizations, or agricultural maps for agencies managing crops. Thus, it is argued that the institutionalization of a land administration system will require the cooperation of the different organizations involved in producing and managing spatial data. The question here is how this inter-organizational cooperation can be achieved so the SDI can be institutionalized. The next section makes a case in favor of adopting a critical approach to address such a question.

2.2.2 Critical Theory and the Study of IT

Within the interpretive approach to information systems, some authors have proposed critical theory as an alternative for guiding researchers. The purpose of critical theory is to reveal the underlying reality of phenomena. Critical theory is associated with the Frankfurt School, a group of German thinkers. The most relevant are: Adorno, Fromm, Habermas, Horkheimer, Marcuse and Weil. Critical theory was founded in the 1920s by Horkheimer [36]. Its epistemology is a mixture of psychoanalysis with Marxism, aiming to emancipate human beings from values of control and efficiency especially from those discourses based on technology [36]. One of the main exponents of critical theory is found in the German philosopher Jurgen Habermas. Habermas [27] argues that science manifests inscribed interests of technological domination yet disguises this fact by appearing to be free from value judgment.

Hence a fundamental idea in the critical study of technology is to reject the positivistic notion that technological artifacts are neutral; i.e., the belief that technology is only a tool of which use and application can be predicted. Technology is the carrier of specific interests and the result of political struggles [9,33,42]. More recently, the philosopher of technology Andrew Feenberg [19,20], a critical theorist, has similarly argued that technological artifacts are not exclusively products of pure engineering and design processes, but are also the result of often conflicting views of the world. Therefore, he proposes that a particular technology cannot be understood without making sense of its specific history, and incorporating local interpretations of its users. Furthermore, Feenberg asserts that the study of technology requires its contextualization—that is, to study its meaning from the point of view not only of the designers but also from users. Moreover, Feenberg argues that this contextualization requires aesthetical and ethical dimensions which are the seeds of emancipation. In the area of organization studies, this approach has been applied by Koch [40] in studying coalitions around technology that are vehicles of organizational change.

In the field of information systems, Lyytinen [48] argues in favor of critical theory as a viable vehicle for conducting information system research (for a thorough review on the state of critical theory applied to the field of IS see Brooke [8]). Lyytinen and Klein propose critical theory as an alternative to the hard science approach, particularly for research on topics such as the adoption and usage of information systems. They argue that information systems should be designed not only to increase efficiency in organizations but "must also increase human understanding and emancipate people from undesirable social and physical constraints, distorted communication and misapplied power" (p. 219). The thrust of the argument is to reveal the social foundations of information systems and to highlight the inappropriateness of undertaking research using engineering approaches exclusively.

For the purposes of this paper I adopt critical theory as the main theoretical lens.* Critical theory is an alternative to the most traditional approaches adopted for studying SDIs that often hinge in the belief that SDIs are neutral.† My critical view of SDI and land administration systems in the context of developing nations will focus on the meaning that local Guatemalan officials assign to those types of systems. I will argue that both are derived from rationales and discourses that are typical of more industrialized nations [17]. It is worth noticing that the technologies required to implement an SDI (i.e., data warehousing, large servers, satellite technologies and digital networks [57]), although amply available in developed nations, may be alien for local officers‡ (i.e., technicians, managers and officers) in developing countries. The availability of these technologies in developed nations reflects their historical and economical context, since their design, production and use entails an abundance of material and human resources [37,38]. The material resources are related to the wealth accumulation of these nations, without which they could not have access to such technologies [17]. It is not only material wealth that is required but also a critical mass of skilled workers, which is a common feature in developed nations [17]. Inter-organizational cooperation is facilitated by centuries of industrialized democracy as is the case in the UK, The Netherlands, Canada and the U.S. Democracy and rule of law are key for institutions to subordinate sectarian interests which unfortunately is not the case in a post-conflict nation such as Guatemala [51,52,53]. For this reason, it is a fair assumption to hold that SDIs are technologies that make more sense in a developed nation (although they may not develop their full potential there; see Steudler et al. [67]), but that may not be the case in a developing country. In order to establish the latter, we need to know the meanings that local users assign to these artifacts.

In sum, this study draws from critical studies of technology two main theoretical assumptions. One is the assumption that SDIs are not neutral as they are the carriers of interests and the result of political struggles. The other assumption holds that the possibility of making sense of how technology is used—that is, from the point of view of researchers and implementers—cannot be achieved without examining both, the history of the technology and its context, and without understanding how the technology is interpreted locally. Accordingly, the case will be analyzed by applying these two

* In the field of IS several researchers have adopted a critical perspective to study IT implementation (see Puri and Sahay [58] who study from a critical perspective the implementation of a GIS in India). Their perspective allowed them to highlight the relevance of local knowledge for implementation and use of the technology. Avgerou [2] presents a review of the literature adopting a critical stance to the study of ICTs in developing countries.

† Exceptions to this approach are the works of Groot [25] and Karikari et al. [37], although they do not directly acknowledge the duality of SDIs in the sense of being enabling and constraining [54].

‡ Local officials in this study were managers, middle managers and technicians working in the different institutions related to land administration in Guatemala. Sometimes I referred to them as users, since they would be the major users of the SDI whenever it is implemented.

assumptions as theoretical lenses to tease meanings out of the data. As indicated above, these two assumptions are derived from the work of Feenberg [19,20], a critical theorist of technology. By articulating these two assumptions it is suggested that critical theory can be applied for conducting IS studies as it points out which particular data to collect and also the relationships among technology, users and power. Consequently, the main goal in interpreting the data will be to synthesize, in the form of themes, the meanings that local users give to SDIs. These themes will be discussed against the backdrop of the two theoretical assumptions; that will facilitate the unraveling of the historical context as well as different sources of power that are related to how local users interpret SDIs.

Once the historical context and discourses have been discussed in the themes, I will propose a course of action for designers and promoters of SDIs intended to address the institutional aspects of implementation. This course of action is structured according to actor-network theory (ANT; see [9,41]) and follows the four moments of translation required to create an ANT [9]. I argue that ANT is an appropriate theoretical framework for articulating this type of course of action given its emphasis on power, meanings and organizations. For example, Callon [10] proposes ANT as a theoretical lens to explain the recalcitrance and strength of economic institutions. In the IS literature ANT has also been applied to explain the institutionalization of systems. For instance, in the UK Bloomfield and colleagues [4,5] have applied ANT to study the institutionalization of new IS related practices in the British NHS. Scandinavian researchers (see, for example, [6,7,29]) have applied ANT to study the institutionalization of health information systems as well as infrastructure standards. Likewise, Silva and Backhouse [65] have applied a theoretical framework that draws on ANT to explain the institutionalization of a system in a research center in Latin America. Of course this proposed course of action should not be taken as a positivistic prediction, i.e., as a "scientific law"; instead it should be considered as an attempt to relate the themes found through interpreting the data with a practical recommendation for those interested in the institutionalization of SDIs. The bringing of ANT provides the technical implications of this paper with a strong theoretical basis. Furthermore, it allows the suggestion of a coherent sequence of actions that practitioners may adopt when implementing similar systems.

2.3 Research Method

To achieve the research objective, I conducted an interpretive study [72]. This is an appropriate approach, as the philosophical underpinnings of an interpretive study presuppose that the meaning of actions is socially constructed [39]. Moreover, an interpretive approach fits the epistemological needs of emphasizing language and interpretations [56]. Especially, it fits the theoretical assumptions presented above in the sense of emphasizing the local, cultural and historical contexts in which the technology is interpreted by its users. Accordingly, bearing in mind the epistemology and ontology of interpretivism, this section describes the series of steps taken in conducting the research.

2.3.1 Data Collection*

Data was collected through interviews, observations and documents. A total of seven interviews were conducted with local officials who work in organizations related to the administration of land in Guatemala. The data was collected throughout a whole week in September of 2004. The institutions visited were CONTIERRA (the organization responsible for mediating conflicts regarding land in Guatemala), the Geographic National Institute (the organization in charge of all cartographic information), UTJ/PROTIERRA (the institution responsible for the national cadastre) and the University of San Carlos, specifically the Faculty of Agronomy (institution in charge of the training of cadastre officials).† The researcher visited these organizations and met with technicians. The interviews were open-ended and the researcher would only formulate straightforward questions regarding the interpretations of the SDI. The main topics of the interviews focused on their use of ICTs (such as GIS, databases and digital maps) and the main problems they face when applying those technologies in their work tasks. Some of the interviews were attended by international consultants who arranged access to the contacts and arranged the logistics of the meetings. Their presence resulted in an added benefit for the researcher, who could appreciate the different interpretations among local officers and international consultants. The researcher took notes during the interviews which later became the basis of the interpretation stage.

The researcher also participated in two major meetings. One was a seminar on the historic context of land ownership in Guatemala. There were three speakers, each presenting a different topic: (1) a historic outline of the problem of land ownership in Guatemala, (2) the possibilities of a land reform and (3) the interpretation of the land ownership problem from the perspective

* I feel confident with the richness of the data. According to Sanders [61] a phenomenological study should be nurtured by three types of data sources: (a) semi-structure interviews, (b) analysis of documents and (c) participants' observations. These three sources provided data for this study. Furthermore, regarding the total interviews Sanders [61, p. 356] indicates: "Quantity should not be confused with quality. The phenomenologist must learn how to engage in in-depth probing of a limited number of individuals. Although the ideal number will vary according to the topic under investigation, too many subjects can become overwhelming. It is realistic to believe that sufficient information may be collected from approximately three to six individuals."

† The list of visited institutions did not pretend to be exhaustive. However, according to the informants these were the most prominent ones in relation to land administration.

of the Mayan people. The other meeting was a workshop organized by the Guatemalan National University, specifically by its school of agriculture. The objective of this workshop was to discuss and establish a research agenda on issues related to land administration systems. This workshop took place over a course of two days and was attended by representatives of all institutions related to the administration of land in Guatemala. The researcher took notes during those meetings and also obtained a copy of the documents summarizing the results of those discussions. The notes of the interviews and those of the meetings were written in Spanish, the native language of the researcher who conducted the fieldwork. The excerpts that are reported in this paper are English translations made by the same researcher.

2.3.2 Interpretation of the Data

Taylor [68, p. 4] considers interpretation as the act of making sense of an incoherent and confusing text: "The interpretation aims to bring to light an underlying coherence or sense." Taylor then concludes that an interpretation should comply with three conditions: (1) the initial text should be evaluated as lacking coherence or sense; (2) the interpretation, that is, the expression that conveys the meaning of the confusing text, should be distinguished from the original text; and (3) the interpretation should make sense to a subject. Taylor [68, p. 5] explains this subject in the following way: "Without such a subject, the choice of criteria of sameness and difference, the choice among the different forms of coherence which can be identified in a given pattern, among the different conceptual fields in which it can be seen, is arbitrary." The third point suggests the hermeneutic circle in which the interpreter goes back and forth, from the text to its interpretation to the subject until finding a satisfactory interpretation, which is agreed upon between the researcher and the subject.

In interpreting the data, I followed Taylor's three conditions. Accordingly, the first step consisted in reading the interview transcripts and field notes with the purpose of identifying parts which seemed perplexing. Once I established that these parts were not idiosyncratic, but rather salient throughout the whole data set, these were synthesized into the initial themes [61]. For example, the "institutional jealousy" theme, discussed in detail in the next section, was identified to be common in most of the interviews. It was also confirmed by the researcher's perception from participating in the aforementioned meetings after conversations with the participants and observations. Once the themes were identified, the second step was to develop an initial interpretation in which the meanings of those themes were proposed, keeping the theoretical lens in mind. The third step consisted in relating the interpretation of the themes to issues that concern the intended audience of this research paper, the academic and practitioner community (the subjects, in the Taylor sense, whom the interpretation is directed to). Therefore, each theme was concluded by a discussion of its theoretical and practical implications.

The overall interpretation was facilitated by the fact that the researcher who collected and interpreted the data was born and raised in Guatemala and by the fact that he has lived in Western Europe and North America for the last twelve years. This provided him with the unarticulated background [68] required to make sense of tacit manifestations of meaning and subtle gestures of Guatemalans as well as those from the international consultants leading the implementations of the SDIs. Moreover, the themes and interpretations presented in this paper are not intended to serve as statistical generalizations [44]. Instead, these are analytical generalizations aimed at enriching the theoretical understanding of the institutionalization of land administration systems.

2.3.3 Context of the Study

The Guatemalan government and Guatemalan society underwent profound transformations in the mid 90s as a consequence of the end of the cold war. We can trace the origins of these transformations as a broad consequence of the peace agreements that ended the Guatemalan civil war in December 1996. With a population of twelve million inhabitants, Guatemala is the most populated country of Central America. Guatemala has been a democratic republic since its independence from Spain in 1821; however, in the 20th century there have been only seven free elections, two in mid century and the latest five since 1986. During the cold war Guatemala was ruled by a military dictatorship supported in part by the USA [23]. Resistance to this regime was organized by a group of rebels allegedly sponsored by the Soviet block that fought a guerrilla war against the government. The conflict lasted for thirty-four years, from 1962 to 1996.

The civil war ended in 1996 with the state of Guatemala agreeing to carry out radical changes in its legislature, constitution and government. The agreement is documented in what is known as the "peace accords." A fundamental component of the *peace accords* is the Guatemalan government's commitment to institute a national cadastre. The cadastre is deemed to be fundamental for solving conflicts originating in disputes regarding land ownership. As will be discussed in detail in this section, land ownership has been one of the major sources of social conflict in Guatemala during the last two centuries [34,50] and has been always at the top in the agendas of issues discussed by grassroots groups. The institutionalization of the national cadastre is believed to be an important step toward solving this long and historical tension in Guatemala. The cadastre would be the first step to legalize and legitimize property rights in areas in which these rights are established only by tradition or are still under dispute.

The efforts to institute the national cadastre in Guatemala have found multiple obstacles (lack of cooperation among political parties, distrust of different stakeholders; see [18]) so it was not until 2005 that the Guatemalan congress approved the cadastre law; without it all the efforts to institutionalize and coordinate different land management agencies would have lacked legal support. The international community has played an important role in promoting the implementation of the national cadastre in Guatemala. Their efforts are aimed at providing funding and technical cooperation. Among the technical cooperation the international community has offered is the implementation of different SDIs which would be the technical platforms on which the national cadastre would operate.

2.4 Results

This section presents the results of the interpretations in the form of themes. In presenting the themes I adopted an iterative technique that is common in interpretive research. This approach consists in going back and forth from the data to theory [24]. Golden-Biddle and Locke dub their technique *telling-showing-telling.* The first *telling* implies an initial reflection on the puzzling aspects of the theme, and by *showing* they mean presenting related data which, in this case, takes the form of quotations from the data. The technique completes its cycle in a second pass through of *telling.* This final *telling* consists in connecting the interpretation of the data with implications for theory and practice. Thus, in the presentation of the themes there is first an introduction of the context (*telling*), followed by data (*showing*) in the form of quotations and the cycle is closed with another *telling* that consists in interpreting the data and reflecting on its implications, both for practice and theory. The quotations presented in the *showing* part represent the themes found in the data, and the reflections on their implications to theory and practice are the result of the author's interpretation of the data in terms of what it means for IS researchers and practitioners.

2.4.1 Theme 1: "Things Here Are Different"

The meaning Guatemalans assign to land management can be summarized in the conversation between a European consultant and a government officer in charge of managing conflicting land disputes. Unfortunately, given that most of the land in Guatemala is owned by a small proportion of the population,* land conflicts are not uncommon [51]. During the aforementioned conversation, the Guatemalan official was describing how most of the conflicts have a common root: that is, landless people occupying land owned by others. She also told the European consultant how, based

* Miethbauer [52] reports that in Guatemala 2.5% of the total 5.3 million farms control 65% of cultivable land.

on Guatemalan law, judges often rule in favor of the owners and order the eviction of the landless.* Then she was interrupted by the European consultant:

> European consultant: "I do not understand. In Europe if someone has two pieces of bread and someone does not have any, a judge can order for the one owning the bread to give one piece to the other."
>
> Guatemalan official (sighing): "Maybe that happens in your country, but this is Guatemala. This is not Europe. Things here are different."

Here we see the clash of two different forms of life: one in which there is a strong belief in the social function of private property, and the other in which the state considers the legality of private property as paramount. This apparent contradiction also suggests the difficulties that international IS consultants may find when attempting to implement systems in developing countries. For an international consultant, such as the one who had the conversation with the Guatemalan official, it is difficult to understand the reason why the institutional background in which they are working impedes system implementation and adoption. Likewise, for the local officers the contradicting interpretations are a source of frustration, because they feel that their international counterparts do not understand their situation.

This is an obstacle to the institutionalization of land administration systems, and other types of systems, especially when those systems are promoted by international agencies. If international consultants do not understand the contradicting meanings that their offered technologies convey, then the technologies will likely be rejected. This clash among different rationales representing different lifestyles is documented in the work of Sahay [60]. He clearly describes how a GIS system developed in North America failed to be institutionalized in Indian organizations. Sahay provides an interpretation of this rejection by pointing out the discourses embedded in the system: discourses of rationality and control of space, which did not make sense for the Indian officials. Thus, in the process of designing information systems and developing plans for implementation, the understanding of the local interpretations of technology is fundamental, if the technology is going to be adopted and eventually institutionalized.

One practical implication of this theme for international IS consultants implementing land administration systems is to be aware of and learn the historical and political context of land management in the particular country [73]. Likewise, local officials should communicate and exercise patience with the international consultants when explaining the differences in their

* Miethbauer [52] also reports that in Guatemala the majority of social conflicts (67%) have roots in land disputes. The same official told us that in 2004 from January to August there had been around 200 land conflicts reported.

country. The more open the communication among users and designers, the better the results will be when designing and implementing information systems [31]. This is a key idea in the alleged effectiveness of applying systems analysis and design methods that emphasize open communication as a means to achieve a shared understanding between users and designers [11].

2.4.2 Theme 2: "We Want Intelligent Systems"

The interpretation of the data shows that while international consultants thought of SDIs as sophisticated technologies, for the Guatemalan officials who deal with the day-to-day aspects of land administration, these technologies were meaningless. For example, when one of these officials was asked about her needs in terms of land administration systems she simply answered: "We can survive without GIS." This is interesting, since in her answer she associates land administration systems with geographic information systems, and not with the sophisticated technology and processes involved in SDIs. Furthermore, she has also taken the opportunity to express her indifference toward those technologies. When asked what kind of system she would like to have, she answered:

> What I want is an intelligent system. For example, if people from the *cadastre* go to two communities, each with a different map, and each map depicts and legitimizes the ownership rights of its corresponding community, I am almost sure that there is going to be conflict. It will be great to have a system that can inform us about these things. Given the violence associated with land conflicts in this country these systems would be a great aid and would really contribute to social stability … Another system that I would like to have is one that can help me to manage the documents of each conflict we handle; we deal with hundreds of those and the documents that are generated by each are very difficult to manage [she showed us a stack of documents associated with one case, and the stack was about 12" high].

We need to take into consideration that the dialogue we were having with the Guatemalan official was about their needs for information and how the different initiatives involving cooperation could help her. In a way, we were asking her about "her Christmas wish list." However, the response suggests her particular interpretation of land management information systems. Her understanding definitively is not the same as the one held by academicians and international consultants. She had a more pragmatic view. While the academic view of land administration systems may possibly be helpful in providing the Guatemalan official with the system she wished, her own reflection on her information needs may suggest that international consultants should include these types of "intelligent systems" as a major part of their discourses, and more importantly as part of their implementation plans.

They will not only be gaining ground in their quest of being understood by their counterparts, but would also be helping relevant and meaningful work tasks, such as preventing and managing land ownership conflicts. By reflecting on this theme, we see that land administration information systems indeed have different levels of understanding. By calling her systems "intelligent," she may infer that those that do not help them are silly.

The gap between the meanings assigned to infrastructures and the pragmatics of doing work with information systems has been clearly identified by those who have applied semiotics [16] as a framework for understanding the relationship between information systems and organizations [45,59,66]. This view regards information systems as semiotic entities. A semiotic entity, like a sentence, has both meaning and syntax. The syntactic part is constituted by the grammar of the language. For example, in the English language there is the rule that the adjective goes before the noun. Nevertheless, as George Orwell pointed out, when criticizing political discourses, a syntactically correct sentence can be meaningless [55]. The meaning of the sentence comes from a hermeneutic interplay between the subject and the text [68]. In seeing SDI as semiotic entities, Stamper and his colleagues would say that the infrastructure is the syntax, as it constitutes an ordered arrangement of computer equipment which may be meaningless for its users. It will gain meaning for its users only when accompanied by "intelligent systems" that can help them to do their work.

A practical implication for IS professionals interested in SDIs is to strive to design not only the implementation of infrastructures, but also the implementation of applications that help officials in their day-to-day tasks. An SDI itself, with all of its technical sophistication, is meaningless for officials who make decisions regarding the actual administration of land. Land administration officials, I argue, will see little benefit in supporting and participating in initiatives that they consider pointless. This line of reasoning is something that has been recognized in the SDI literature. Groot [25, p. 292] observes that SDI should be focused on helping the work tasks of end users; otherwise, he says, "… the complexity will tend to drive the development into an academic or impractical direction."

In terms of the institutionalization of a land administration system, it follows that the support of local officials will be fundamental. Yet institutionalization will not occur if the system presented is meaningless to them. Conversely, local officials would support the institutionalization of a system that aims to solve their problems. Hence international agencies planning to implement a land administration system in a developing country should begin by designing and developing systems bottom up—that is, to take into consideration the needs of information of individual institutions involved and design systems that would help them in supporting their day-to-day tasks. This is relevant since it was found that the impression that most local officials had of a land administration system was the implementation of geographic information systems supported by satellites; for local officials those were not such intelligent systems, despite what the international consultants thought.

2.4.3 Theme 3: "Institutional Jealousy"

As mentioned previously, the proper operation of a land administration system requires data from multiple sources [57]. These sources are different organizations such as ministries, municipalities or regional authorities [35]. It is widely accepted that without cooperation and coordination among the organizations involved, land administration information systems can neither be designed nor implemented [37,70]. Achieving cooperation among institutions is very difficult in post-conflict nations such as Guatemala, wherein many of its institutions are undergoing radical reforms. Such is the case with the institution that, as stipulated in Guatemalan law, is responsible for managing the geographical matters of the country. This institution, so we were told, has had five directors in a very short period of time and most of its specialists are hired on a temporary basis:

> We cannot do research because we are 029s (temporary contracts). However, we have heard that one ministry is developing their own maps with their own technology. But we cannot tell you more because we are not invited to the meetings when they discuss this. Organizations keep their maps and data. There is a great deal of institutional jealousy.

Institutional jealousy, understood as the tendency of organizations to withhold information and neglect cooperation with other agencies, is a key obstacle for the institutionalization of land administration systems. Those in charge of designing and implementing such systems should be aware of this tendency of non-cooperation among organizations. The challenge for them is twofold: either motivate organizations to begin sharing and cooperating with each other, or design systems that do not require institutional cooperation. We deem the latter as unfeasible, given the complexity of land information systems and the multiple levels of data that these types of systems require to operate. The first alternative, although difficult, seems to be the only viable one. The question, then, is how to achieve cooperation among institutions when this has not been the norm for centuries.

The European Union, recognizing the criticality of inter-organizational cooperation for the institutionalization of an SDI, created a body in 1993 called EUROGI (EURopean umbrella Organization for Geographic Information). A body of a similar nature would be fundamental for implementing SDIs in post-conflict nations like Guatemala in which institutions exhibit institutional jealousy.* Furthermore, from the point of view of a critical theorist, it can be argued that the coordinating body would need to promote open communication, aiming at an ideal speech situation [27] in which individuals from different organizations can exchange their thoughts and fears sincerely.

* Obviously this recommendation has as its basis that it has worked for Europeans and is based in the theory of communication of Habermas. Our recommendation is based on considering that "institutional jealousy" is grounded on fears exacerbated by the unstable environment of a post-conflict nation.

As will be discussed in the next section, without an open and sincere dialogue institutional jealousy may not be overcome.

2.4.4 Theme 4: "We Are Used to Life without Law"

As discussed above, the notion of land administration systems is derived from rational normative discourses [25, p. 37] that presuppose a world in which the state functions properly, in the sense that each of its institutions operates faithfully to its prescribed role. Although not completely perfect, this view may be the case in developed nations such as those in North America and some parts of Europe whereby the rationalization of bureaucracy is much more institutionalized than in incipient democracies in the third world. In the case of technology transfer, difficulties arise when the ordered institutional setting of a developed nation is taken for granted and thought of as a given in a developing country.

For example, Williamson and Ting [76], in discussing what is required to re-engineer a land administration system, quite appropriately highlight the need to make institutional changes, without mentioning exactly how to achieve such changes:

> Legislative reform appropriately follows policy development, technological advances and institutional reform and should be a support process in re-engineering land administration systems.

Discussing the Guatemala case in the context of this theme will show that the above normative statement, which is written in cold and neat logic, will be meaningless considering its historical context and lack of operational content.

A quick examination of the history of Guatemala will reveal decades of dictatorship, civil war and social and political unrest which have created obstacles to the consolidation of democratic institutions [23,49,53]. The director of a major cadastre project expressed it this way:

> Here it is very difficult to establish a land administration system because that would mean stopping abuses from land owners. They obviously are going to resist it because we have lived without the rule of law for a very long time. We were used to life without law regulating land ownership.

Inequities of land ownership in Guatemala can be traced to colonial times in the sixteenth century, and continuing after the independence of Guatemala from Spain in 1821 [49]. This inequality was accentuated at the end of the nineteenth century when the military-led liberal government decided to bestow incentives for those landowners cultivating coffee [34] as well as the decision of government officials in the early twentieth century to grant huge estates to their friends for the same purposes [50]. This created severe political and social tensions in the middle of the twentieth century that eventually led to the implementation of land reform by virtue of a decree from the democratically elected government of Jacobo Arbenz Guzmán [26]. This land reform expropriated large farms and estates, and without a doubt was one of the main events that ultimately triggered the overthrow of Arbenz from government [21,23,62].

Thus, the linkage between power and land ownership has permeated Guatemalan politics and institutions, and poses a daunting challenge for those attempting to implement local SDIs for land management. Their efforts may be regarded as a threat to those who have been in a position of power. A simple re-engineering approach would not work.

What for an international engineer is considered to be a neutral technology of which the ultimate goal is to impact development [25] for a local officer such a technology may be interpreted as an instrument that could be a source of conflict. Here, the observation made by Feenberg discussed in the theory section becomes relevant. SDIs will remain alien for their users in contexts such as that of Guatemala if designers and implementers do not take into account local interpretations. Therefore, engineers and consultants should not only be skillful on the technical aspects of SDIs but also should be aware of the meaning that their users may assign to those technologies.

In the next section a line of action is proposed that may help designers and implementers of SDIs overcome some of the difficulties addressed in each of the themes above.

2.5 A Proposed Model for Approaching Institutionalization

The institutionalization of a land administration information system is, I argue, an issue that has to do with power. It is a power issue, given that each institution does not see any incentive in sharing its information [15]. The organizational members do not see any material benefit for sharing information. Moreover, it was found that they believe that if they were to give away information, they may be regarded as dispensable. Likewise, there is a lack of disciplinary techniques ensuring that organizations would comply with sharing their information [14]. Thus, the challenge for the designers would be to facilitate cooperation among the diverse organizations, regardless of their believing the system is a threat and their certainty that they can get away without cooperating [46]. This is a formidable task considering that institutionalized practices of non-cooperation and institutional jealousy, as implied above, have been in place for centuries. The breaking of this organization inertia [28] would require the exercise of power at two different levels: meanings and techniques of discipline [12].

Exercise of power at the level of meanings refers to the land administration designers' task of persuading different organizations that the land administration system is something from which they are going to obtain a benefit. Techniques of discipline, on the other hand, refer to techniques that will ensure cooperation. For example, the cadastre law that is currently being discussed by the Guatemalan congress at the writing of this manuscript would be the regulatory framework for establishing those disciplinary techniques. A theoretical framework that incorporates these two dimensions of power—meaning and techniques of discipline—is ANT (see [9,41,43] for prime examples of the application of these theories).

In terms of ANT, the challenge for those proposing the land administration system will be to define, for those institutions that are supposed to adopt them, what ANT theoreticians call obligatory passage points (OPP). To explain this concept, let us use Dahl's [13] terminology for a power relation. In a power relation, Dahl explains that there are two different agencies—the "As" and the "Bs"; an "A" exercises power over a "B" when "A" makes "B" do something "B" would otherwise not do. In the case of land administration systems, the "As" would be those who want to design and implement the land administration system and the "Bs" would be those opposing it. Thus, "A" would like to set the land administration system as an OPP for "B." OPPs are the result of "translations" after which "B" has no other choice but to accept the OPP as created by "A."

There are four "moments" of translation: *problematization, interessement, enrolment,* and *mobilization.* Callon [9] explains these "moments" in detail, and in this section I will elaborate on their application as guidelines for understanding the political and institutional aspects of the design and implementation of land administration systems. Table 2.1 shows the relationship between the proposed tactical recommendations and the themes that emerged from the data.

2.5.1 Problematization

The first step of translation is *problematization* or how to become indispensable. *Problematization* occurs when one actor is given a problem and through rhetorical means presents the solution of that problem in terms of his or her resources. In this way, one group of actors defines an OPP. In the context of implementing an SDI, the task for the designers (either international or local consultants) is to portray the problems of land administration clearly and explain how the SDIs are the solution to those problems. Moreover, the designers must identify the actors required to be enrolled; i.e., they need to identify who are the "Bs" of their power relation. As discussed above, this is not an easy task, given the large number of organizations that need to work together and the differences in interests among different stakeholders. However, without this, it is argued that institutions and officials will not fully cooperate with the implementation of SDIs. At this point, the theme of

TABLE 2.1

The Relationship among Themes and the Tactical Model

Themes	Relationship with Tactical Model
(1) "Things here are different"	
If international consultants are unaware of the context and history of the country, initiatives toward implementing SDIs are doomed to be ignored.	Any attempt to *interest* local officers by international consultants should be based on the latter understanding the values and interests of the former.
(2) "We want intelligent systems"	
SDIs need to be planned bottom up—that is, first solving relevant problems in institutions before SDIs are presented as sophisticated mega projects.	*Problematization* is not going to occur if the SDI is perceived as something detached from day-to-day tasks of local officers. SDIs need to be presented as solutions to current and relevant problems.
(3) "Institutional jealousy"	
SDIs will require the sharing of information and coordination among institutions. This is going to be difficult if members of one organization distrust members of other organizations, since they may feel that their jobs and positions may be threatened.	The institutions involved in an SDI are not going to play their respective *roles* of cooperation required for an SDI to operate if there is distrust and if their members feel insecure about their jobs and positions in their organizations. Organizational members will neither be *enrolled* nor *mobilized* in the presence of institutional jealousy.
(4) "We are used to living without law"	
SDIs will require a strong legal frame to define the obligations and rights of organizations and stakeholders. However, in a country not used to the rule of law, intense negotiations and strengthening of democratic agencies such as courts and congress will be required.	The *problematization* of SDI, the *interessement* of different actors, as well as their *enrolment* and *mobilization*, will not occur in the absence of a strong legal frame.

"intelligent systems" becomes relevant, as local officials would like to participate and support systems that they see would solve their problems. In other words, presenting the SDI with technological jargon and without addressing practical and tangible problems of the institutions will not suffice for a successful *problematization.*

2.5.2 Interessement

The second step of translation is called *interessement.* After the identities of the actors and OPPs have been defined and established during *problematization,* the group of actors experiencing the problem (i.e., the "Bs") must commit to the solution. This is a huge challenge for designers and promoters of SDIs, as most of them do not have the legal and disciplinary resources to obtain a commitment from the authorities assuring that they will cooperate and implement the SDI. The signing of the 1996 peace accords, which contemplated the implementation of a land administration system, was an event that could be classified as an effort toward achieving *interessement*. Moreover, if the cadastre law that is currently being discussed in the Guatemalan congress is finally approved [18], then a large step toward a successful *interessement* will be achieved. The *interessement* of the Guatemalan government regarding the implementation of the SDI has been difficult. For example, an international government that initially was funding the implementation of the SDI decided to withdraw their financial support until the Guatemalan government approves the cadastre law. In the context of ANT, this is a clear power move on behalf of the international government to achieve the *interessement* of the Guatemalan government.

2.5.3 Enrolment

The third step of translation is *enrolment*. During this step, alliances among different actors are consolidated through bargaining and making concessions. Once the *problematization* and *interessement* have been successful, the *enrolment* as presented by Callon consists in defining the exact role of each actor. In the case of the SDI this will correspond to an operations plan indicating exactly the role of each institution in the implementation of the SDI. In such a plan of operations, the "As" of the power relation should detail not only the courses of actions that the "Bs" must follow, but also the benefits that the latter will obtain by agreeing to play their roles. In implementing an SDI, the "A" of the power relation could be a coordinating body such as EUROGI that will be bringing together other institutions; those would be the "Bs" of the relation. Enrolment will not occur until the bargaining and negotiations for successful compliance by "B" are completed; i.e., the implementation and operations plan should contemplate dialogue and discussions among "As" and "Bs" so their roles, as well as their obligations and rights, are clearly defined. Without finalizing that, the *enrolment* of the institutions will not occur.

2.5.4 Mobilization

The fourth and final step of translation is the *mobilization* of the allies. According to Callon, *mobilization* implies that actors will become spokespersons of the groups they claim to represent. In the case of the implementation of the land administration system, *mobilization* will be the actual implementation of the plan. During this stage, keeping the alliances stable will be the challenge for the designers and promoters of the system. This stability entails that each representative participating in the process should be considered as legitimate by their constituencies. This is achieved, according to Callon, by actors remaining faithful to the terms of the *enrolment* and to the constituencies they represent. *Mobilization* is a fragile "moment" in the steps of translation because alliances may fall apart if the organizations do not obtain the offered benefits or if any of the other achieved stages of translation, i.e., *problematization, interessement* or *enrolment,* were to fail. For example, a local official who initially was supporting and participating in the design and implementation of the SDI may abandon the program if she believes that the offered "intelligent systems" that were supposed to ease her work tasks are not going to be delivered after all. Therefore, *mobilization* calls for those managing the SDI to monitor constantly the state of the alliances and the status of delivering that which was promised.

2.5.5 Reflection on the Frailty of the Translation Process

Of course, nothing is guaranteed, since the resources of the "Bs" can be superior to those of the "As." Some organizations may see their interests threatened by the implementation of a land administration system and may mobilize their resources to stop it. The above proposal seems appropriate as it suggests how to institutionalize a particular practice. Its application constitutes a tactical and strategic approach to SDI implementation, and it is claimed that it is a fresh one, given that most of the literature on SDIs takes a normative stance regarding the cooperation of institutions; that is, the literature indicates that institutional changes are required [35,37,52], but does not tell us how to achieve them. The suggested tactical approach to SDIs consists in accepting that the SDIs are not neutral and can originate contradicting interpretations between designers and users. Thus, the first step in this tactical approach would consist in keeping an open mind to understand local interpretations. The second step would be to plan a set of activities aimed at clarify meanings and conciliate different interests. For such a purpose the moments of translation of ANT can be a useful template.

However, ANT cannot be taken in a normative fashion. Its nature will require designers and promoters of land administration systems to interpret local situations, interests and tactics. Indeed, one of the limitations of ANT is that it is always difficult to identify the limits of the network, where they begin and where they end.* Furthermore, this model can only be applied when the actors and institutions involved are clearly identified. As suggested above, its application does not guarantee success by any means and does not take care of the limitations of the theory. However, ignoring the political nature of changing institutions would almost guarantee failure. Therefore a proper design of an SDI should include not only all of the technological components, but also a proposed course of action aimed at generating cooperation among institutions so institutionalization is achieved.

* For a comprehensive critique of ANT see Collins and Yearly (1992) and Ihde (1990).

2.6 Conclusions

This paper has contributed to the current body of knowledge by conducting a critical study of the implementation of an SDI in a post-conflict developing nation. Such a critical approach allowed us to identify concrete institutional themes that will need to be addressed if SDI designers and promoters want to be successful in their endeavors. In addition, this research constitutes a contribution to the traditionally normative approaches to SDIs' implementations. Instead of merely indicating that institutional cooperation is a fundamental condition for successful SDIs' implementations, this paper concentrates on the reasons local officers point out as barriers for cooperation among institutions. It has also put forward a tactical model—based on ANT—that may help practitioners in planning for institutionalization. Even though this proposal may be qualified as unfeasible given the institutionalization of practices associated with the four themes found, and the power imbalances that still prevail in a post-conflict nation such as Guatemala, it is argued that the proposal is a contribution given that it is not a normative one. It operates at a tactical level so that practitioners may use the proposed model as a guideline for implementation, particularly when formulating strategies and tactics for overcoming political aspects of implementing SDIs. They cannot do it without interpreting the world view of the locals.

The results of this study may also be useful for international agencies that are promoting and sponsoring local SDIs in countries like Guatemala. Before embarking on their tasks of cooperation, international consultants and engineers should study the historical context of land administration in the country they are going to work for. In addition, engineers and designers should be aware of the institutional setting in which the SDIs are being proposed. Being aware of the historical context and institutional setting of the proposed SDIs will enhance their job as engineers and at the same time will add to their work a more human and ethical dimension. In addition, their work will be more effective since this study has shown that institutionalization cannot be achieved without attention to local interpretations and particular interests.

One of the limitations of critical research is that it does not propose concrete actions to solve more permanent power relations [1]. In a way, this study exhibits that exact limitation, which it has addressed by proposing the ANT model for implementation. Although ANT is usually applied in descriptive studies, I have articulated a more proactive application which may help implementers and designers not only in the way described above, but also as an instrument for monitoring the state of institutional coordination. Another limitation of the research is that it concentrates on general obstacles of SDIs and land administration; in the fieldwork these were talked about as general terms. Because of that, it is not possible to know if, given the opportunity, local officials would have assigned different meanings to different components of an SDI. The same can be said for the fact that the research concentrated exclusively on Guatemala; it is not possible to know what the interpretation of local officials of other countries would be in different situations. Thus, an area of further research could be to study the meanings local officials assign to different parts of the SDI, as well as expanding the study to other nations so comparisons across countries can be conducted.

All in all, the implementation and adoption of complex technologies such as SDIs offer formidable challenges not only to designers but also for adopters. The difficulty of this challenge is exacerbated when designers and adopters belong to different cultures and come from different political and economical systems. We have known for many years that designers need to be sensible and appreciative of cultural differences [3,11,73]. Yet what has been found in this research is that such a sensitivity needs to be accompanied by patience and by a savvy political approach, based, in turn, much more on consensus and bargaining than on conflict, and that operates at a tactical level. For adopters of SDIs in a post-conflict nation, the challenge is daunting since they have to deal not only with their own convoluted political and social context, but also have to make an effort to understand what their international counterparts are telling them. Besides, given their different interpretations of SDIs, local officials should make an effort in order to be understood by their counterparts. In sum, a critical study such as this is a contribution to the overall efforts toward effective communication among the actors involved in the implementation of an SDI. Open communication is fundamental to clarify contradicting interpretations which in turn is fundamental to achieving cooperation among institutions. Designers and promoters of SDIs may bear that in mind considering that institutionalization will not occur merely by decree. As shown in this case, institutionalization of SDIs will not occur just because of the will of a government expressed in accords and decrees; institutionalization will be rather the result of effective and conscious political action.

References

1. M. Alvesson and S. Deetz, *Doing Critical Management Research,* London: SAGE, 2000.
2. C. Avgerou, "Recognising alternative rationalities in the deployment of information systems," *Electronic Journal of Information Systems in Developing Countries,* vol. 3, no. 7, 2000, pp. 1–15.
3. D.E. Avison and A.T. Wood-Harper, *Multiview: An Exploration in Information Systems Development,* London: Blackwell Scientific Publications, 1990.
4. B.P. Bloomfield and A. Best, "Management consultants: Systems development, power and the translation of problems," *The Sociological Review,* vol. 40, no. 3, 1992b, pp. 533–560.

5. B.P. Bloomfield and R. Coombs, "Information technology, control and power: The centralization and decentralization debate revisited," *Journal of Management Studies,* vol. 29, no. 4, 1992, pp. 459–484.
6. J. Braa and C. Hedberg, "The struggle for district-based health information systems in South Africa," *Information Society,* vol. 18, no. 2, 2002, pp. 113–127.
7. J. Braa, E. Monteiro, and S. Sahay, "Networks of action: Sustainable health information systems across developing countries," *MIS Quarterly,* vol. 28, no. 3, 2004, pp. 337–362.
8. C. Brooke, "Editorial: Critical research in information systems: Issue 2," *Journal of Information Technology,* vol. 17, no. 4, 2002, p. 179.
9. M. Callon, "Some elements of a sociology of translation: Domestication of the scallops and the fishermen of St Brieuc Bay," in John Law, ed., *Power, Action and Belief,* Vol. Sociological Review Monograph 32, The Sociological Review, London: Routledge & Kegan Paul, 1986, pp. 196–233.
10. M. Callon, "Techno-economic networks and irreversibility," in John Law, ed., *A Sociology of Monsters: Essays on Power, Technology and Domination,* Vol. 38, Sociological Review Monograph, London: Routledge, 1991, pp. 132–161.
11. P.B. Checkland and J. Scholes, *Soft Systems Methodology in Action,* Chichester: Wiley, 1990.
12. S.R. Clegg, *Frameworks of Power,* London: SAGE Publications, 1989.
13. R. Dahl, "The concept of power," *Behavioural Science,* vol. 2, 1957, pp. 201–205.
14. C. Dandeker, *Surveillance, Power and Modernity: Bureaucracy and Discipline from 1700 to the Present Day,* Cambridge: Polity Press, 1990.
15. T.H. Davenport, R.G. Eccles, and L. Prusak, "Information politics," *Sloan Management Review,* Fall 1992, pp. 53–65.
16. U. Eco, *A Theory of Semiotics,* in J. Young and P. Walton, ed., *Critical Social Studies,* London: The Macmillan Press Ltd, 1976.
17. A. Escobar, *Encountering Development,* N. B. Dirks and G. Eley, ed., Princeton, NJ: Princeton University Press, 1995.
18. R. Estrada, Varias razones por las que aún no se aprueba la Ley de Catastro. 2005, El Periodico, Guatemala.
19. A. Feenberg, *Questioning Technology,* New York: Routledge, 1999.
20. A. Feenberg, *Transforming Technology: A Critical Theory Revisited,* 2nd ed., New York: Oxford University Press, 2002.
21. C. Forster, *The Time of Freedom: Campesino Workers in Guatemala's October Revolution,* Pittsburgh, PA: University of Pittsburgh Press, 2001.
22. A. Giddens, *The Constitution of Society,* Cambridge: Polity Press, 1984.
23. P. Gleijeses, *Shattered Hope: The Guatemalan Revolution and the United States, 1944–1954,* Princeton, New Jersey: Princeton University Press, 1991.
24. K. Golden-Biddle and K.D. Locke, *Composing Qualitative Research,* Toronto: SAGE Publications, 1997.
25. R. Groot, "Spatial data infrastructure (SDI) for sustainable land management," *ITC Journal,* vol. 3, no. 4, 1997, pp. 287–294.
26. Guatemalan Government, *Decreto Numero 900: Ley de Reforma Agragaria,* Presidencia de la Republica de Guatemala, 1952.
27. J. Habermas, *Knowledge and Human Interests,* Boston: Beacon Press, 1972.
28. M.T. Hannan and J. Freeman, "Structural inertia and organizational change," *American Sociological Review,* vol. 49, 1984, pp. 149–164.

29. O. Hanseth and K. Braa, "Hunting for the treasure at the end of the rainbow. Standardisation of corporate IT infrastructure," *Computer Supported Cooperative Work,* vol. 10, no. 3/4, 2001, pp. 261–292.
30. O. Hanseth, E. Monteiro, and M. Hatling, "Developing information infrastructure standards: The tension between standardisation and flexibility," *Science, Technology and Human Values,* vol. 21, no. 4, 1996, pp. 407–426.
31. R. Hirschheim and H.K. Klein, "Realizing emancipatory principles in information systems development: The case for ETHICS," *MIS Quarterly,* vol. 1994, March 1994, pp. 83–109.
32. M.D. Iatau, "An introduction to the use of a case study methodology to review cadastral reform in Papua, New Guinea," in *Proceedings of the Commission 7 Symposium, 64th Permanent Committee of the International Federation of Surveyors,* Singapore, 1997, p. 9.
33. D. Ihde, *Technology and the Lifeworld: From Garden to Earth,* The Indiana Series in the Philosophy of Technology, ed. Don Ihde, Bloomington: Indiana University Press, 1990.
34. R.H. Jackson, ed., *Liberals, the Church, and Indian Peasants: Corporate Lands and the Challenge of Reform in Nineteenth-Century Spanish America,* Albuquerque: University of New Mexico Press, 1997.
35. S. Jacoby, J. Smith, L. Ting, and I. Williamson, "Developing a common spatial data infrastructure between state and local government: An Australian case study," *International Journal of Geographical Information Science,* vol. 16, no. 4, 2002, pp. 305–322.
36. A.G. Johnson, *The Blackwell Dictionary of Sociology,* Oxford: Blackwell, 1995.
37. I. Karikari, J. Stillwell, and S. Carver, "Land administration and GIS: The case of Ghana," *Progress in Development Studies,* vol. 3, no. 3, 2003, pp. 223–242.
38. I. Karikari, J. Stillwell, and S. Carver, "The application of GIS in the lands sector of a developing country: Challenges facing land administrators in Ghana," *International Journal of Geographical Information Science,* vol. 19, no. 3, 2005, pp. 343–362.
39. H. Kincaid, "Social sciences," in Peter Machamer and Michael Silberstein, ed., *The Blackwell Guide to the Philosophy of Science,* Malden, Massachussetts: Blackwell Publishers, 2002, pp. 290–311.
40. C. Koch, "Building coalitions in an era of technological change," *Journal of Organizational Change Management,* vol. 13, no. 3, 2000, pp. 275–288.
41. B. Latour, "The powers of association," in John Law, ed., *Power, Action and Belief: A New Sociology of Knowledge,* Vol. Sociological Review Monograph 32, London: Routledge & Kegan Paul, 1986, pp. 264–280.
42. B. Latour, *We Have Never Been Modern,* London: Harvester Wheatsheaf, 1993.
43. J. Law, "On the methods of long-distance control: Vessels, navigation and the Portuguese route to India," in John Law, ed., *Power, Action and Belief,* Vol. Sociological Review Monograph 32, The Sociological Review, London: Routledge & Kegan Paul, 1986, pp. 234–263.
44. A.S. Lee and R.L. Baskerville, "Generalizing generalizability in information systems research," *Information Systems Research,* vol. 14, no. 3, 2003, pp. 221–243.
45. J. Liebenau and J. Backhouse, *Understanding Information: An Introduction,* Macmillan Information Systems Series, ed. Ian O. Angell, London: Macmillan, 1990.

46. S. Lukes, *Power: A Radical View,* Studies in Sociology, ed. Anthony Giddens, London: The Macmillan Press Ltd, 1974.
47. K. Lyytinen, "Information systems and critical theory," in M. Alvesson and H. Willmott, ed., *Critical Management Studies,* London: SAGE, 1992, pp. 158–169.
48. K.J. Lyytinen and H.K. Klein, "The critical theory of Jurgen Habermas as a basis for a theory of information systems," in Enid Mumford, Rudy Hirschheim, Guy Fitzgerald, and Trevor Wood-Harper, ed., *Research Methods in Information Systems,* Amsterdam: North Holland, 1985, pp. 219–236.
49. S. Martínez, *La Patria del Criollo: Ensayo de interpretación de la realidad colonial guatemalteca,* México: Fondo de Cultura Económica, 1998.
50. D. McCreery, "Wage labor, free labor, and vagrancy laws: The transition to capitalism in Guatemala, 1920–1945," in William Roseberry, Lowell Gudmundson, and Mario Samper Kutschbach, ed., *Coffee, Society, and Power in Latin America,* Baltimore: The Johns Hopkins University Press, 1995, pp. 206–231.
51. T. Melville and M. Melville, *Guatemala: The Politics of Land Ownership,* New York: Free Press, 1971.
52. T. Miethbauer, "Social conflict and land tenure institutions as a problem of order policy the case of Guatemala," in *Proceedings of Deutscher Tropentag,* Berlin, 1999.
53. H. Miller, "Liberal modernization and religious corporate property in nineteenth-century Guatemala," in Robert H. Jackson, ed., *Liberals, the Church, and Indian Peasants,* Albuquerque: University of New Mexico Press, 1997, pp. 95–122.
54. W.J. Orlikowski, "Knowing in practice: Enacting a collective capability in distributed organizing," *Organization Science,* vol. 13, no. 3, 2002, pp. 249–273.
55. G. Orwell, *All Propaganda is Lies,* London: Secker and Warburg, 1998.
56. C. Pavitt, *The Philosophy of Science and Communication Theory,* Huntington, New York: Nova Science Publishers, Inc., 2000.
57. A. Phillips, I. Williamson, and C. Ezigabalike, "Spatial data infrastructure concepts," *The Australian Surveyor,* vol. 44, no. 1, 1999, pp. 20–28.
58. S.K. Puri and S. Sahay, "Participation through communicative action: A case study of GIS for addressing land/water development in India," *Information Technology for Development,* vol. 10, no. 3, 2003, pp. 179–199.
59. R. Sabherwal and Y.E. Chan, "Alignment between business and IS strategies: A study of prospectors, analyzers, and defenders," *Information Systems Research,* vol. 12, no. 1, 2001, pp. 11–33.
60. S. Sahay, "Implementation of information technology: A time–space perspective," *Organization Studies,* vol. 18, no. 2, 1997, pp. 229–261.
61. P. Sanders, "Phenomenology: A new way of viewing organizational research," *Academy of Management Review,* vol. 7, no. 3, 1982, pp. 353–360.
62. S. Schlesinger and S. Kinzer, *Bitter Fruit: The Untold Story of the American Coup in Guatemala,* London: Anchor Books, Doubleday, 1982.
63. W.R. Scott, *Institutions and Organizations, Foundations for Organizational Science,* ed. Gillian Dickens, London: SAGE Publications, 1995.
64. L. Silva, "Theoretical approaches for researching power and information systems: The benefit of a Machiavellian view," in Debra Howcroft and Eileen Trauth, ed., *Handbook of Critical Information Systems Research: Theory and Application,* London: Edward Elgar Publishing, Inc., 2005, pp. 47–69.

65. L. Silva and J. Backhouse, "The circuits-of-power framework for studying power in institutionalization of information systems," *Journal of the Association for Information Systems,* vol. 4, no. 6, 2003, pp. 294–336.
66. R. Stamper, "The semiotic framework for information systems research," in Hans-Erik Nissen, H.K. Klein, and R. Hirschheim, ed., *Information Systems Research: Contemporary Approaches and Emergent Traditions,* Amsterdam: North Holland, 1991, pp. 515–527.
67. D. Steudler, A. Rajabifard, and I. Williamson, *Evaluation of Land Administration Systems,* 2004. Retrieved from http://www.sli.unimelb.edu.au/research/SDI research/publications/files/SteudlerEtal2004.pdf
68. C. Taylor, "Interpretation and the sciences of man," *Review of Metaphysics,* vol. 25, no. 71, 1971, pp. 3–51.
69. J. Turkstra, N. Amemiya, and J. Murgia, "Local spatial data infrastructure, Trujillo-Peru," *Habitat International,* vol. 27, no. 4, 2003, p. 669.
70. P. van der Molen, "Institutional aspects of 3D cadastres," *Computers, Environment & Urban Systems,* vol. 27, no. 4, 2003, p. 383.
71. M. Wade and J.S. Hulland, "The resource-based view and information systems research: Review, extension, and suggestions for future research," *MIS Quarterly,* vol. 28, no. 1, 2004, pp. 107–142.
72. G. Walsham, *Interpreting Information Systems in Organizations,* John Wiley series in information systems, ed. Richard Boland and Rudy Hirschheim, Chichester: John Wiley, 1993.
73. G. Walsham, V. Symons, and T. Waema, "Information systems as social systems: Implications for developing countries," in S.C. Bhatnagar and N Bjorn-Andersen, ed., *Information Technology in Developing Countries,* North-Holland, 1990, pp. 51–61.
74. I. Williamson, "Land administration 'best practice': Providing the infrastructure for land policy implementation," 2001. Retrieved from http://www.sli.unimelb.edu.au/subjects/451/418/418 2001/Best Practice.htm
75. I. Williamson and G. Mathieson, "The Bangkok land information system project—Designing an integrated land information system for a large city in the developing world," *Canadian Institute of Surveying and Mapping Journal,* vol. 46, no. 2, 1992, pp. 153–164.
76. I. Williamson and L. Ting, "Land administration and cadastral trends—A framework for re-engineering," 2003.

3

Integrating Spatial Information and Business Processes: The Role of Organizational Structures

Ezra Dessers, Geert Van Hootegem, Joep Crompvoets, and Paul H. J. Hendriks

CONTENTS

3.1 Introduction

Spatial data infrastructures (SDIs) have been developed in many countries worldwide and are now accepted as an essential infrastructure in modern society (Crompvoets, Rajabifard, et al. 2008). However, a plethora of disparate approaches regarding definitions, objectives, assessments, and analyses of SDIs can be found in the literature. Because SDI is such a multifaceted, dynamic, and complex concept (Grus, Crompvoets, and Bregt 2007), some form of complexity reduction is needed in order to deliver a meaningful contribution to the SDI research. The focus of this chapter is defined from the viewpoint of the organizations, which are key stakeholders of the SDI. The relevance of this perspective derives from the fact that the actual objective of an SDI is not to serve the data handling functions per se, but to serve the needs of the user community (Rajabifard, Feeney, and Williamson 2002).

Despite the many definitions and approaches, an SDI is typically defined as a set of interacting resources for facilitating and coordinating spatial data access, use, and exchange (Rajabifard et al. 2002; Nedović-Budić, Pinto, and Budhathoki 2008). Thus, an SDI deals with data and, eventually, information flows. Organization theory stresses that information does not flow in the void, but instead runs through a network of business processes, and that the relevance of the information flows depends on their value for these business processes (Daft 2001). In fact, data are turned into information through the business processes. Connecting SDIs to business processes therefore provides a viable and promising option for SDI analysis (Chan and Williamson 1999).

We use the term "business process" in a broad sense, referring to processes in any organized form of cooperation and not just in the restricted sense of commercial activities by private sector organizations. A business process conceived in this broad sense is the way in which organizations create products and services (Daft 2001; Desmidt and Heene 2005). It usually takes the form of a sequence of interrelated activities, which turn a certain input of resources into an output of products or services. From an organizational viewpoint, spatial information flows and the business processes involved should be analyzed together (Vandenbroucke et al. 2009). They are intertwined, as is their performance (de Sitter 2000).

The key question is, then, which factors might impact the performance of both business processes and their spatial information flows. This chapter puts the focus on organizational structures without denying that legal, technological, and other aspects also have an influence on performance (Crompvoets, Bouckaert, et al. 2008; Vandenbroucke et al. 2009). As will be substantiated in this chapter, organizational structures have to be identified as a crucial yet undervalued element.

The aims of this chapter are to (1) identify concepts for describing the relation between organizational structures and the integration of spatial information in the business processes, and (2) explore empirically the applicability of these concepts in order to further shape the research agenda.

The chapter is divided into four parts. In this introduction, we established the organizational viewpoint. In the second part, a social systems (Luhmann 1984; de Sitter, den Hertog, and Dankbaar 1997) approach is used to gain insight into the relation between organizational structures and the performance of business processes and their information flows. In the third part, we investigate the applicability of this approach in an exploratory case study. In the fourth part, we conclude with general remarks and discuss the application of the concepts developed in this chapter in current SDI research.

3.2 Organizational Structures and Business Process Performance

The introduction of this chapter suggested that spatial information flows take place in a network of business processes. In this part we introduce basic concepts to identify and describe organizational structures that are relevant in explaining the integration of business processes and spatial information flows. This part is divided into three sections. The first argues that organizational structures are the key factor affecting business process performance. Functional concentration is introduced as the central concept for analyzing organizational structures. The second section argues that environmental demands are crucial factors for assessing the effectiveness of organizational structures. In the third section, we address the impact of organizational structures on the performance of business processes and their spatial information flows.

3.2.1 Organizational Structures and Functional Concentration

Simons (2005) argues that organizational structures impact the performance of every individual in an organization and are therefore the most important determinant of performance. Although other dimensions, such as individual behavior, organization culture, or social relations, may be important to make business processes successful, a systems view on organizations (Daft 2001; Seddon 2008; Achterbergh and Vriens 2009) regards organizational structures as the fundament on which the other dimensions rest.

An organizational structure can be defined as inter-related ways in which (1) labor is divided into distinct tasks and (2) coordination is achieved among these tasks (Mintzberg 1993). This results in (1) a production structure and (2) a control (or coordination) structure (de Sitter et al. 1997). As Figure 3.1 shows, the production structure can be defined as the architecture of grouping and linking production, preparation, and support functions in relation with the business process flows. The control structure can be defined as the architecture of grouping and linking the coordination and regulation activities (de Sitter 2000). The term "control" is used in organization design literature in a neutral sense to indicate the function of regulation and it does not

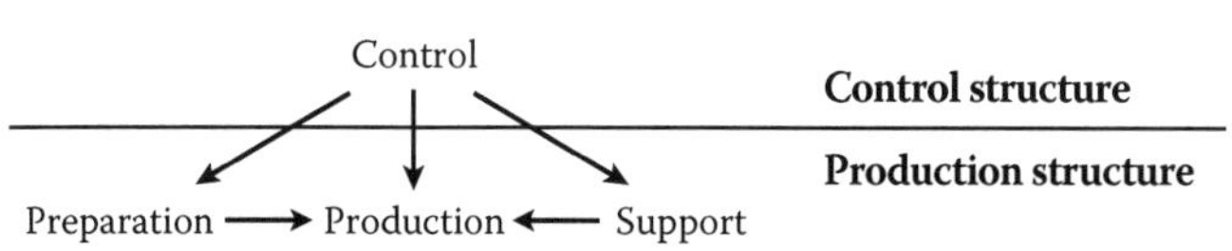

FIGURE 3.1
The control and the production structure.

imply any particular form (e.g., both direct command and facilitation qualify as control).

With regard to the production structure, de Sitter (2000) argues that the division of labor is essentially based on decisions regarding the level of functional concentration, which concerns the grouping of similar functions into the same organizational subsystems. Grouping most of the spatial data-related functions of an organization in a large GIS department is an example of functional concentration. A far reaching integration of GIS functions (like database updating) into the business processes of the various departments is an example of functional deconcentration. Based on this criterion, two basic structuring forms can be distinguished: a function-based and a process-based division of labor.

A function-based division of labor implies a high level of functional concentration. Similar or related activities are brought together in one organizational unit (Mintzberg 1993; Stoner, Freeman, and Gilbert 1995). The creation of the product or service is divided across various units, which all produce only a limited part of the complete product or service. A process-based division of labor builds on a low level of functional concentration (which could be identified as deconcentration). All activities related to the production and marketing of a product or service (or a related group of products or services) or related to a certain type of customer are brought together in one organizational unit (Stoner et al. 1995). The business processes are assembled in parallel, heterogeneous, and relatively autonomous units (Van Hootegem 2000). Both basic forms can be found within and between numerous organizations. Evidently, hybrid forms can be identified also (Buelens 1992; Mintzberg 1993; Dawson 1996; Hatch 1997; Van Hootegem et al. 2008).

With regard to the control structure, again two basic alternatives are possible: centralized and decentralized (Van Hootegem 2000; Daft 2001). In the centralized control structure, the coordination function is separated from the productive, preparative, and support functions. The coordination capacity is then situated at a management level, away from the operational work. If, on the contrary, the coordination function is integrated in the production structure, a decentralized control structure is created. In this chapter, separating the coordination function from the executing functions is also regarded as a form of functional concentration, as suggested by Van Hootegem (2000). The level of functional concentration of the control function is high in the case of a centralized control structure. In a decentralized control structure, the level of functional concentration of the control function is low.

3.2.2 Business Processes: Environment and Performance

The previous section described how choices regarding the functional concentration typically imply different forms of organizational structures.

The question remains as to what factors define the effectiveness of a certain organizational structure. This section argues that an organization is a relatively open system, which interacts with its environment (van Eijnatten 1993; Daft 2001; Desmidt and Heene 2005). The environment can be described in terms of the external demands that are put on the business processes, like demands for efficiency or reliability (Bekkers 1998). These demands are defined by the claims of customers, suppliers, competitors, and regulatory bodies like governments.

3.2.2.1 Organizational Structures and Business Process Performance

Coordination mechanisms can be considered as the glue holding organizational structures together (Mintzberg 1993). Environmental demands are relevant factors for assessing the effectiveness of organizational structures because they influence the amount of coordination needed to make the business processes perform well (Daft 2001). The modern sociotechnical systems (MSTS) approach offers a social systems-based framework (Luhmann 1984) for the analysis and design of organizations (van Eijnatten 1993; de Sitter 2000; van Amelsvoort 2000). Although this chapter is primarily based on sociotechnical theory as developed in The Netherlands, the term "sociotechnical systems" was originally coined in the 1960s by Trist and Bamforth (1951) in the United Kingdom, and the theory was further developed and applied in North America, Australia, and Scandinavia (for an international comparison, see van Eijnatten 1993).

MSTS suggests that the organizational structure creates the necessary and boundary conditions to meet the environmental demands. Based on Ashby's law of requisite variety (Ashby 1957), MSTS states that the coordinating capacity of an organizational unit should be in proportion to its coordinating needs. A function-based division of labor can be very effective in dealing with efficiency demands because internal efficiency of every business process step can be maximized. When the environment is relatively stable, the coordination needs in the production structure are usually limited because the routine tasks can be standardized and centrally coordinated (Daft 2001). However, when the environment becomes more dynamic and flexibility and innovation gain importance, the business processes should be able to react swiftly to environmental changes. The need for coordination increases.

A function-based organizational structure is likely to have trouble in meeting these needs: On the one hand, each unit controls only a very small step in the entire business process; on the other hand, every unit has to deal with many business processes and their connected demands. Structures that are organized in process-oriented ways run parallel with each other and cause less interference. Coordination problems could be minimized by placing division of the work into as many subtasks as possible with logically connected streams of tasks (Van Hootegem et al. 2008). It should be noted that several other design theories have many points in common with MSTS, including business process reengineering (BPR), as described by Hammer and Champy (1993), and Lean thinking (Womack, Jones, and Roos 1990).

3.2.2.2 Environmental Changes

The longstanding prosperity of not only the private but also the public sector was based on the far reaching function-based division of labor (Meier and Hill 2005; Du Gay 2000). However, environmental developments have challenged this model, both in the private (Hammer and Champy 1993) and in the public (Osborne and Gaebler 1992) sectors. These environmental changes can be characterized by (1) an increasing level of uncertainty, causing the environment to become more complex to handle (Pfeffer and Salancik 1978; Daft 2001), and (2) a rising level of dependence on external resources, leading to a growing network of interdependencies between organizations. In particular, the streams of information are ever greater in number and size (Castells and Himanen 2002; Roche and Caron 2004). As a result of this evolution, business processes are confronted with increasing demands for flexibility and innovation (Bolwijn and Kumpe 1991; Van Hootegem et al. 2008). They become less predictable, routine, and transparent.

3.2.3 Spatial Information Flows

The use and exchange of information has become increasingly important in contemporary society (Castells and Himanen 2002; Roche and Caron 2004). Both in number and size, the information flows are always growing. These flows often have a spatial component (Longhorn and Blakemore 2008). The growing importance of (spatial) information itself raises the need for mutual alignment and cooperation between and within organizations (Campbell and Masser 1995; Omran 2007). At the same time, SDI initiatives aspire to support organizations and their business processes in dealing with environmental demands (Rajabifard et al. 2002).

Based on the line of reasoning that was developed in the previous sections of this chapter, we assume that organizational structures could play an important role in the integration of spatial information flows and business processes. However, a systematic account of how different structures relate to business process performance and how spatial information flows are to fit in these appears lacking in SDI literature.

An increasing stream of studies addresses the organizational sides to SDI (Nedović-Budić, Pinto, and Warnecke 2004; Masser 2006; Koerten 2008; Craglia and Campagna 2009). When considering organizational factors that influence SDI development, topics like strategy, leadership, and technology get significant scholarly and managerial attention. Little attention has been paid so far to the characteristics of the business processes and organizational structures through which spatial data access, use, and exchange take place. Most publications stress the importance of (organizational) culture and (individual and organizational) behavior (Coleman, Groot, and McLaughlin 2000; Nedović-Budić and Pinto 2001; Wehn de Montalvo 2003; Craig 2005; Masser 2005; de Man 2007; Omran 2007; Koerten 2008; Nedović-Budić et al. 2008; Van Loenen and van Rij 2008).

However, several references to business processes and organizational structures can be found in the assorted SDI writings, especially in the context of interorganizational interoperability (Nedović-Budić and Pinto 2001). It is apparent that international assessments of SDIs, like the Inspire State of Play (Vandenbroucke and Janssen 2008), and SDI readiness studies (Delgado Fernández et al. 2006) contain a number of organizational elements, like the existence of a coordinating body or the level of participation of data users and producers in the initiative.

Rajabifard (2003) studied the engagement of states in a regional SDI initiative and also dealt with the possible utility of his approach in analyzing the participation of organizations in actual spatial data sharing. Nedović-Budić, Feeney, et al. (2004) examined the structural characteristics of interaction mechanisms between organizations, concluding that spatial data sharing efforts involve redefinition of existing tasks and structures and the establishment of new ones. McDougall, Rajabifard, and Williamson (2007) analyzed the characteristics of spatial data sharing partnerships between state and local governments in Australia from an organizational perspective encompassing the role and place of the individual organization in the partnership. Harvey and Tulloch (2006) stated that SDI development requires stronger connections between SDI policy and other government policies and activities. They recommended that data sharing should be part of the business processes at all levels of government.

Some studies focus on the business processes of individual organizations. Based on organizational change literature, Pornon (2004) pointed out that the utility of spatial information for organizations should be seen from a systems approach encompassing diversification of business flows and a primary emphasis on the significance of information for these business flows. Based on a literature review, Bekkers and Moody (2006) distinguished between instrumental and institutional factors that influence the usage of spatial information for policy making. The social network analysis by Omran and Van Etten (2007) revealed that a hierarchical organization structure could put serious constraints on spatial data sharing. Based on mixed-method research, Vonk, Geertman, and Schot (2007) concluded that knowledge about business processes and knowledge about technology should be brought together. Finally, only a few studies take the viewpoint of a specific business process or policy field to assess SDI performance, like poverty management (Akinyemi 2007) or local planning (Nedović-Budić, Pinto et al. 2004).

3.2.4 Summary

We argued that the integration of spatial information flows and business processes is expected to be influenced by the structural characteristics of the organizations involved. Functional concentration was proposed as a central concept for describing these organizational structures. In the next part of this chapter we present an exploratory case study in which the ability of the concept of functional concentration to discern between different forms of organizational structures is explored.

3.3 An Exploratory Case Study Approach

In order to gain a better grasp of the process and issues related to the integration of spatial information in business processes, we conducted an exploratory case study (Yin 2003). The aims of the study were to (1) investigate the ability of the concept of functional concentration to detect differences in organizational structures, and (2) lay down criteria for developing a research design for investigating the relation between organizational structures and spatial information practices. No particular propositions or hypotheses were tested.

3.3.1 Case Selection

For the exploration pursued in this study, we defined a case as a public sector organization. We focused on the public sector for three reasons:

- SDI initiatives are generally launched by public-sector organizations (Masser 2005).
- Most information in the public sector could be linked to a location (Longhorn and Blakemore 2008). Therefore, the potential benefits of SDI in the public sector can be enormous (Masser, Rajabifard, and Williamson 2007).
- The availability of clear and precise government information is of great importance to organizations outside the public sector, as well as to individual citizens (Stiglitz 1999; Longhorn and Blakemore 2008).

For reasons of comparability, two cases were selected from a group of organizations with similar tasks and responsibilities (i.e., among the five provincial administrations in the region of Flanders, Belgium). Provincial administrations are typical examples of public-sector organizations covering the wide range of responsibilities of a certain governmental level. The cases are considered to be representative of the provincial governments in Flanders. While we sought similarity in the overall organizational environment and functions, we chose two provincial administrations that appeared to be mutually the most divergent with regard to spatial information use and spatial information policy: Limburg and West-Vlaanderen. Limburg is situated in eastern Flanders, which is the northern region of Belgium. West-Vlaanderen is located in the west of Flanders. Some general features of both provinces are presented in Table 3.1.

Case study data were collected through survey data (Crompvoets et al. 2009) and in-depth interviews with the GIS coordinator, an executive official, and the manager of the spatial planning department of both provincial governments. The GIS coordinator mainly talked about the spatial information use and the spatial information policy of the organization. The executive official provided information on the organizational structure. The manager of the spatial planning department was interviewed to explore the integration of spatial information within specific business processes, notably spatial planning processes.

The case study comprised three steps:

1. We made a general overview of the differences in spatial information use and spatial information policy between both administrations.
2. We investigated the applicability of the theoretical concepts in describing the organizational structure of both cases.
3. The information gathered was analyzed and evaluated in order to lay down criteria for developing a research design for further in-depth and comparative research.

3.3.2 Exploring Spatial Information Use and Policy

Spatial information use and policy are used in this study as a rough indicator at the organizational level for assessing the overall integration of spatial information flows in the business processes. In order to explore the differences between two provincial governments in spatial information use and policy, we relied on both survey and interview data. As shown in Table 3.2, West-Vlaanderen has a higher level of data production and sharing, compared to Limburg, and a higher percentage of its employees make use of

TABLE 3.1

The Provinces of Limburg and West-Vlaanderen

	Limburg	West-Vlaanderen
Area (km^2)	2,422	3,144
Population (2008)	826,690	1,150,487
Capital	Hasselt	Bruges
Number of employees of provincial administration	1,150	800

TABLE 3.2

Exploring Spatial Information Use by and Policy of the Provincial Administrations of Limburg and West-Vlaanderen

Criteria	Limburg	West-Vlaanderen	Source
Data production	Low	High	Survey
Data sharing	Low	High	Survey
Percentage of spatial information users	10	20	Survey
Strategic plan	No	Yes	Interviews
Support local governments	No	Yes	Interviews
Focus of GI coordination	Internal	Both internal and external	Interviews
Integration in spatial planning processes	Limited	Extensive	Interviews
Province as knowledge center	Marginal issue	Core issue	Interviews

spatial information. Limburg has no strategic spatial information policy plan and does not develop initiatives to support the local governments within their territories. West-Vlaanderen does have a strategic plan (Provincie West-Vlaanderen 2005) and has a history of supporting the local governments and cooperating with them. The focus of the GI coordination in Limburg is largely internal; for West-Vlaanderen, the external coordination with other governments and organizations is just as important.

Interviews with the heads of the spatial planning departments in both organizations revealed that the integration of spatial information within the spatial planning processes is more extensive in West-Vlaanderen than in Limburg. This was measured by comparing the actual level of spatial data access, use, and exchange within every step of the spatial planning process for both provinces. Finally, West-Vlaanderen considers becoming a knowledge center for external people and organizations as its core mission, while this is not as relevant in the case of Limburg. To conclude, this explorative comparison shows major differences between the provincial governments in their spatial information use and policy.

3.3.3 Organizational Structures

3.3.3.1 The Case of Limburg

An organization chart is a diagram that depicts the overall structure of an organization. The organization chart of Limburg in Figure 3.2 shows that a mainly function-based production structure can be identified, with five separate departments for each policy field. Within each department, a further function-based division is applied. For instance, the Department of Infrastructure, Spatial Planning, Environment and Nature

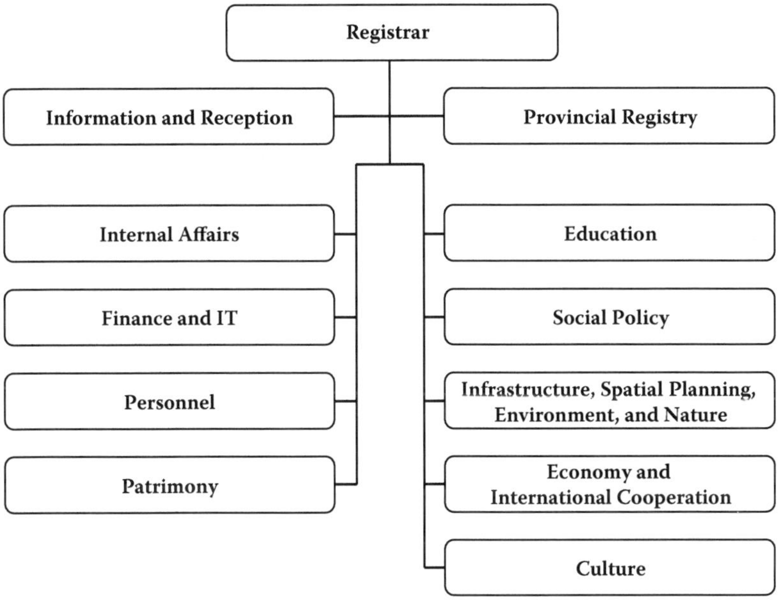

FIGURE 3.2
Internal organization chart of the provincial administration of Limburg (excluding the political bodies and the federal and regional departments).

has three divisions: (1) Infrastructure, consisting again of the sections of Administration, Roads, and Waterways and Domains; (2) Spatial Planning, containing the sections of Permits, Planning and Policy, and Mobility; and (3) Environment and Nature, composed of the sections of Permits, and Planning and Policy.

The support and preparative functions are mainly concentrated in four function-based departments. A limited number of (mostly administrative) support functions are situated at the various secretariats within the policy departments. Within Finance and IT, a two-person section is responsible for GIS coordination, database management, and support, while the actual use of spatial information takes place within the different thematic departments and sections.

Like most of the organizations applying a far reaching function-based labor division (de Sitter 2000), the province of Limburg has a centralized, hierarchical coordination structure. Although interdepartmental consultation is encouraged, most coordination is done through the hierarchical channels. Limburg comes close to the ideal type of a function-based organization. Interestingly, the province has recently designed a new organization chart, in which the functional concentration will be strengthened. Generally speaking, a new hierarchical level will be created, bundling the existing policy departments into two "superdepartments." At the same time, some support functions that at the moment are still situated within policy departments, such as school building management in the Department of Education, will be moved to a support department.

3.3.3.2 The Case of West-Vlaanderen

The organization chart in Figure 3.3 shows that the province of West-Vlaanderen is structured according to five policy domains and four management domains. A policy domain is the grouping of similar policy matters and contains units that deliver some form of external output. The policy domain "living environment," for example, contains units responsible for matters such as spatial planning, mobility, infrastructure, environment, and nature. Coordination teams are set up between and within policy domains to answer the need for consultation.

A management domain is the grouping of connected management activities and contains units that are responsible for specific management affairs, such as finance. A management domain is oriented toward the performance and the management of the organization as a whole. The four management domains are (1) government, (2) planning, (3) realization, and (4) support. The organization is configured as a matrix in which the policy domains are on the vertical axis and the management domains on the horizontal axis. Within every policy and management domain, different configurations of preparatory, support, and/or management staff and executing units can be found.

This production structure could be described as a move from a function-based organization toward the hybrid form of a matrix organization, which typically tries to combine vertical functional departments with horizontal

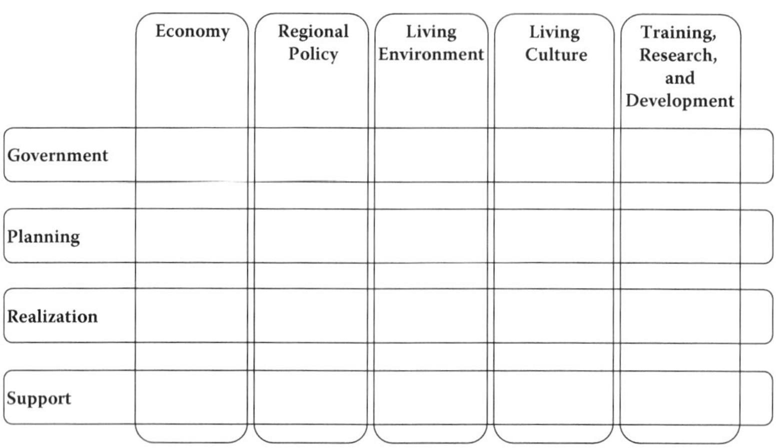

FIGURE 3.3
Internal organization chart of the provincial administration of West-Vlaanderen (excluding the political bodies and the federal and regional departments).

processes or projects. At the level of the whole organization, West-Vlaanderen tries to combine horizontal process-based functions, as well as support and control functions, and vertical, largely functional but also regional policy and knowledge center perspectives. Although the general support units, such as personnel, still exist, the province strives to integrate support and preparative functions with production functions in the policy domains. The GIS coordination is situated within the horizontal support domain. As in Limburg, the application of GIS is integrated into the various policy domains.

The province of West-Vlaanderen acknowledges the need for a specific, integrated policy toward each of the different regions of the province, but the responsibility for these regional policies is entrusted to a separate policy domain. The policy domain of "training, research, and development" accounts for the new, core mission of the province as knowledge center for external customers across the different policy domains. Again, this integrated mission is allocated to a separate policy domain. In short, an integrated process or customer orientation is taken into account, but is realized mainly by separate policy domains and through consultation forums.

As for the control structure, the number of hierarchical levels was drastically reduced by the introduction of the matrix organizational chart in 2007, and direct consultation and coordination teams were created to assure process and customer orientation. For example, the head of the spatial planning unit reports directly to the political level; in the former structure, he had three hierarchical levels above him. His performance assessment is done by the chairman of the coordination team of his policy domain. Some domains do have an extra hierarchical level, but never more than one. The horizontal domains are more process oriented, and their heads make up the management team that steers the overall organizational course.

Despite the matrix-style organization chart, the provincial administration is still largely made up of specialized, function-based units. However, a tendency toward a less hierarchic control structure and more direct mutual coordination in order to improve process and customer orientation was found. Moreover, the policy domains of "regional policy" and "training, research, and development" could be regarded as steps toward a more process-based approach.

3.3.4 Lessons Learned for Research Design

The concept of functional concentration allows us to identify differences in the organizational structures of both provincial administrations. Limburg is still very bureaucratic in nature, combining high levels of functional concentration and centralization. West-Vlaanderen, on the other hand, is searching for ways to evolve in the direction of process-oriented and less centralized organization. Moreover, the concept of functional concentration allows for interpretation of the organization charts in terms of the actual production and control structures. The matrix chart of the province of West-Vlaanderen is not entirely the result of striving toward functional deconcentration; it is also the outcome of a simple reshuffling of existing divisions in a new scheme.

Based on the theoretical argument presented earlier in this chapter, these differences in organizational structure may partly be causally related to the observed variations in spatial information use and policy (Table 3.2). Given the level of environmental complexity and dynamism, the theoretical argument suggests that a move away from bureaucratic and centralized divisions of labor could contribute to a high-performing integration of spatial information flows in the business processes. The case study findings seem to confirm that a high level of functional concentration correlates with lower scores for spatial information use and policy, as can be seen in Limburg. Conversely, West-Vlaanderen seems to combine a lower level of functional concentration with higher scores for spatial information use and policy.

However, the exploratory study reveals a number of methodological weaknesses, which should be resolved in order to reach a satisfying design for further research:

- A general assessment of an organization's spatial information use and policy might be useful as a first indication, but it is too general to serve as a dependent variable in comparative research. Analyzing the integration of spatial information flows in the business processes will require a more detailed analysis of specific business processes within an organization. Not all business processes take the same course. Some may stay largely within the borders of one organizational unit, while others involve different steps through multiple units. Also, the environmental demands can differ: In spite of the general evolution toward more flexibility, this is not necessarily the case for every individual business process. Besides, the relevance of spatial information can vary considerably between business processes.
- Linking the overall structure of an organization to the integration of spatial information with its business processes is probably too broad a scope for empirical research. This is especially the case when studying an organization that is active in a broad spectrum of policy domains, such as a provincial administration. Indications were found in the interviews that the internal structure of the various divisions within the same organization can be relatively different.
- Many of the business processes in which spatial information is involved cross the boundaries of a single organization (Bekkers and

> Moody 2006; McDougall et al. 2007). These value chains imply an interorganizational structure (Gereffi et al. 2005). The interorganizational division of labor can be achieved by function or by process (de Sitter 2000). Allocating different steps of the business process to different organizations by bundling similar functions (such as land survey) in specialized organizations results in a function-based interorganizational division of labor. Conversely, by making each organization responsible for complete business processes, a process-based division of labor is achieved. The interorganizational structure refers to the allocation of different steps in the business process to different organizations and the way in which coordination between the steps is organized.

To summarize, only a systematic and comparative analysis of different business processes within and between various organizations may lead to significant findings regarding the relation between organizational structures and the integration of spatial information in the business processes.

3.4 Conclusion

This chapter has considered a number of organizational issues relating to spatial information flows. Based on theoretical considerations, the level of functional concentration was proposed as the key parameter of organizational structures impacting the integration of spatial information flows in the business processes. The exploratory case study showed how the concept of functional concentration may be used to distinguish between different forms of organizational structures. The findings of this study suggest that concrete business processes, both within and across the boundaries of the organization, should be the primary subject of further comparative research.

In the multidisciplinary case studies that will be performed in the context of the SPATIALIST project (Crompvoets, Bouckaert, et al. 2008), a case is therefore defined as a business process between and within government organizations in the region of Flanders, Belgium, in which spatial information is accessed, used, and exchanged. Four cases are selected: (1) spatial zoning planning, (2) flood planning, (3) address management, and (4) traffic accident registration. Within each case, about six organizations are selected as embedded cases (Yin 2003). Apart from the various disciplinary analyses (such as the study of the impact of organizational structures), the disciplinary approaches will be combined in a multidisciplinary analysis. The examination of different combinations of causally relevant elements should lead to the identification of distinct "recipes" for high-performing integration of spatial information flows in business processes.

References

Achterbergh, J., and D. Vriens. 2009. *Organizations. Social systems conducting experiments*. New York: Springer.

Akinyemi, F. O. 2007. Spatial data needs for poverty management. In *Research and theory in advancing spatial data infrastructure concepts*, ed. H. Onsrud, 261–277. Redlands, CA: ESRI Press.

Ashby, W. R. 1957. *An introduction to cybernetics*. London: Chapman and Hall.

Bekkers, V. 1998. Grenzeloze overheid. *Over informatisering en grensveranderingen in het openbaar bestuur*. Alphen aan de Rijn: Samson (in Dutch).

Bekkers, V., and R. Moody. 2006. Geographical information and the policy formulation process: The emergence of a reversed mixed scanning mode? In *Information and communication technology and public sector innovation*, ed. V. Bekkers, H. van Duivenboden, and M. Thaens, 103–120. Amsterdam: IOS Press.

Bolwijn, P., and T. Kumpe. 1991. *Marktgericht ondernemen: Management van continuïteit en vernieuwing*. Assen: Van Gorcum/Stichting Management Studies (in Dutch).

Buelens, M. 1992. *Management en effectieve organisatie*. Tielt: Lannoo/Scriptum (in Dutch).

Campbell, H., and I. Masser. 1995. *GIS and organizations. How effective are GIS in practice?* London: Taylor & Francis.

Castells, M., and P. Himanen. 2002. *The information society and the welfare state*. Oxford, England: Oxford University Press.

Chan, T. O., and I. P. Williamson. 1999. Spatial data infrastructure management: Lessons from corporate GIS development. Paper presented at AURISA 99, Blue Mountains, New South Wales.

Coleman, D., R. Groot, and J. McLaughlin. 2000. Human resource issues in the emerging GDI environment. In *Geospatial data infrastructure. Concepts, cases and good practice*, ed. R. Groot and J. McLaughlin, 236–244. Oxford, England: Oxford University Press.

Craglia, M., and M. Campagna. 2009. Executive summary. In *Advanced regional spatial data infrastructures in Europe*, ed. M. Craglia and M. Campagna, 10. Luxembourg: Office for Official Publications of the European Communities.

Craig, W. J. 2005. White knights of spatial data infrastructure: The role and motivation of key individuals. *URISA Journal* 16:5–13.

Crompvoets, J., G. Bouckaert, G. Vancauwenberghe, D. Vandenbroucke, J. Van Orshoven, K. Janssen, J. Dumortier, et al. 2008. Interdisciplinary research project: SPATIALIST. Spatial data infrastructures and public sector innovation in Flanders (Belgium). Paper presented at GSDI-10 Conference, St. Augustine, Trinidad.

Crompvoets, J., E. Dessers, T. Geudens, K. Janssen, G. Vancauwenberghe, D. Vandenbroucke, and M. Van hoogenbemt. 2009. *Het GDI-netwerk in Vlaanderen. Een kwantitatieve verkenning van het gebruik en de uitwisseling van geodata in Vlaanderen*. Leuven: Spatialist, K. U.Leuven.

Crompvoets, J., A. Rajabifard, B. Van Loenen, and T. Delgado Fernández. 2008. Introduction. In *A multi-view framework to assess spatial data infrastructures*, ed. J. Crompvoets, A. Rajabifard, B. Van Loenen, and T. Delgado Fernández, 1–6. Melbourne, Australia: Melbourne University Press.

Daft, R. L. 2001. *Organization theory and design.* Mason, OH: Southwestern College Pub.

Dawson, S. 1996. *Analyzing organizations.* New York: Palgrave.

Delgado Fernández, T., K. Lance, M. Buck, and H. Onsrud. 2006. Assessing an SDI readiness index. Paper presented at *From Pharaohs to Geoinformatics,* Caïro, Egypt.

de Man, E. 2007. Are spatial data infrastructures special? In *Research and theory in advancing spatial data infrastructure concepts,* ed. H. Onsrud, 33–54. Redlands, CA: ESRI Press.

de Sitter, L. U. 2000. *Synergetisch produceren: Human resources mobilisation in de produktie: Een inleiding in structuurbouw.* Assen: Van Gorcum (in Dutch).

de Sitter, L. U., J. F. den Hertog, and B. Dankbaar. 1997. From complex organizations with simple jobs to simple organizations with complex jobs. *Human Relations* 50:497–534.

Desmidt, S., and A. Heene. 2005. *Strategie en organisatie van publieke organisaties.* Tielt: Lannoo (in Dutch).

Du Gay, P. 2000. *In praise of bureaucracy. Weber, organization, ethics.* London: SAGE Publications.

Gereffi, G., J. Humphrey, and T. Sturgeon. 2005. The governance of global value chains. *Review of International Political Economy* 12:78–104.

Grus, L., J. Crompvoets, and A. B. Bregt. 2007. Multi-view SDI assessment framework. *International Journal of Spatial Data Infrastructures Research* 2:32–53.

Hammer, M., and J. Champy. 1993. *Reengineering the corporation—A manifesto for business revolution.* New York: HarperCollins Publishers.

Harvey, F., and D. Tulloch. 2006. Local-government data sharing: Evaluating the foundations of spatial data infrastructures. *International Journal of Geographic Information Science* 20:743–768.

Hatch, M. J. 1997. *Organization theory. Modern, symbolic and post-modern perspectives.* Oxford, England: Oxford University Press.

Koerten, H. 2008. Assessing organisational aspects of SDIs. In *A multi-view framework to assess spatial data infrastructures,* ed. J. Crompvoets, A. Rajabifard, B. Van Loenen, and T. Delgado Fernández, 235–254. Melbourne, Australia: Melbourne University Press.

Longhorn, R. A., and M. Blakemore. 2008. *Geographic information. Value, pricing, production and consumption.* Boca Raton, FL: Taylor & Francis.

Luhmann, N. 1984. *Soziale systeme. Grundriss einer algemeinen theorie.* Frankfurt: Suhrkampf (in German).

Masser, I. 2005. *GIS worlds. Creating spatial data infrastructures.* Redlands, CA: ESRI Press.

———. 2006. What's special about SDI related research? *International Journal of Spatial Data Infrastructures Research* 1:14–23.

Masser, I., A. Rajabifard, and I. P. Williamson. 2007. Spatially enabling governments through SDI implementation. *International Journal of Geographic Information Science* 21:1–16.

McDougall, K., A. Rajabifard, and I. P. Williamson. 2007. A mixed-method approach for evaluating spatial data sharing partnerships for spatial data infrastructure development. In *Research and theory in advancing spatial data infrastructure concepts,* ed. H. Onsrud, 55–73. Redlands, CA: ESRI Press.

Meier, K. J., and G. C. Hill. 2005. Bureaucracy in the twenty-first century. In *The Oxford handbook of public management,* ed. E. Ferlie, L. E. Lynn, and C. Pollitt, 51–71. Oxford, England: Oxford University Press.

Mintzberg, H. 1993. *Structure in fives. Designing effective organizations.* Englewood Cliffs, NJ: Prentice Hall.

Nedović-Budić, Z., M.-E. F. Feeney, A. Rajabifard, and I. P. Williamson. 2004. Are SDIs serving the needs of local planning? Case study of Victoria, Australia, and Illinois, USA. *Computers, Environment and Urban Systems* 28:329–351.

Nedović-Budić, Z., and J. K. Pinto. 2001. Organizational (soft) GIS interoperability: Lessons from the U.S. *International Journal of Applied Earth Observation and Geoinformation* 3:290–298.

Nedović-Budić, Z., J. K. Pinto, and N. R. Budhathoki. 2008. SDI effectiveness from the user perspective. In *A multi-view framework to assess spatial data infrastructures,* ed. J. Crompvoets, A. Rajabifard, B. Van Loenen, and T. Delgado Fernández, 273–303. Melbourne, Australia: Melbourne University Press.

Nedović-Budić, Z., J. K. Pinto, and L. Warnecke. 2004. GIS database development and exchange: Interaction mechanisms and motivations. *Journal of the Urban and Regional Information Systems Association* 16:15–29.

Omran, E.-S. E. 2007. *Spatial data sharing: From theory to practice.* Enschede: Wageningen Universiteit.

Omran, E.-S. E., and J. Van Etten. 2007. Spatial data sharing: Applying social network analysis to study individual and collective behavior. *International Journal of Geographical Information Science* 21:699–714.

Osborne, D., and T. Gaebler. 1992. *Reinventing government—How the entrepreneurial spirit is transforming the public sector.* New York: Plume.

Pfeffer, J. S., and G. R. Salancik. 1978. *The external control of organizations: A resource dependence perspective.* New York: Harper and Row.

Pornon, H. 2004. Ingénierie des SIG: comment les SIG entrent dans les organisations. In *Aspects organisationnels des SIG,* ed. S. Roche and C. Caron, 148–167. Paris: Hermes/Lavoisier (in French).

Provincie West-Vlaanderen. 2005. *GIS West II. Strategisch plan 2005–2010.* Brugge: Provincie West-Vlaanderen.

Rajabifard, A. 2003. SDI diffusion—A regional case study with relevance to other levels. In *Developing spatial data infrastructures. From concept to reality,* ed. I. P. Williamson, A. Rajabifard and M.-E. F. Feeney, 78–94. London: Taylor & Francis.

Rajabifard, A., M.-E. F. Feeney, and I. P. Williamson. 2002. Future directions for SDI development. *International Journal of Applied Earth Observation and Geoinformation* 4:11–22.

Roche, S., and C. Caron. 2004. Introduction. In *Aspects organisationnels des SIG,* ed. S. Roche and C. Caron, 17–22. Paris: Hermes/Lavoisier (in French).

Seddon, J. 2008. *Systems thinking in the public sector.* Axminster: Triarchy Press.

Simons, R. 2005. *Levers of organization design.* Harvard: Harvard Business School Press.

Stiglitz, J. E. 1999. Knowledge as a global public good. In *Global public goods,* ed. I. Kaul, I. Grunberg, and M. Stern, 308–326. Oxford, England: Oxford University Press.

Stoner, J. A. F., E. R. Freeman, and D. R. Gilbert. 1995. *Management,* 6th ed. Englewood Cliffs, NJ: Prentice Hall Inc.

Trist, E. L., and K. W. Bamforth. 1951. Some social and psychological consequences of the longwall method of coal getting. *Human Relations* 4:3–38.

van Amelsvoort, P. 2000. *The design of work and organisation. The modern sociotechnical systems approach.* Vlijmen: ST-Groep.

Vandenbroucke, D., J. Crompvoets, G. Vancauwenberghe, E. Dessers, and J. Van Orshoven. 2009. A network perspective on spatial data infrastructures: Application to the sub-national SDI of Flanders (Belgium). *Transactions in GIS* 13:105–122.

Vandenbroucke, D., and K. Janssen. 2008. *Spatial data infrastructures in Europe: State of play 2007.* Leuven: K. U. Leuven.

van Eijnatten, F. M. 1993. *The paradigm that changed the work place.* Assen: Van Gorcum.

Van Hootegem, G. 2000. *De draaglijke traagheid van het management: Tendensen in het productie-en personeelsbeleid.* Leuven: Acco (in Dutch).

Van Hootegem, G., P. van Amelsvoort, G. Van Beek, and R. Huys. 2008. *Anders organiseren & beter werken. Handboek sociale innovatie en verandermanagement.* Leuven: Acco (in Dutch).

Van Loenen, B., and E. van Rij. 2008. Assessment of spatial data infrastructures from an organizational perspective. In *A multi-view framework to assess spatial data infrastructures,* ed. J. Crompvoets, A. Rajabifard, B. Van Loenen, and T. Delgado Fernández, 173–192. Melbourne, Australia: University of Melbourne Press.

Vonk, G., S. Geertman, and P. Schot. 2007. New technologies stuck in old hierarchies: The diffusion of geo-information technologies in Dutch public organizations. *Public Administration Review* 67:746–756.

Wehn de Montalvo, U. 2003. In search of rigorous models for policy-oriented research: A behavioral approach to spatial data sharing. *URISA Journal* 15:19–28.

Womack, J. P., D. T. Jones, and D. Roos. 1990. *The machine that changed the world.* Ontario, Canada: Collier-MacMillan.

Yin, R. K. 2003. *Case study research. Design and methods.* London: Sage Publications.

4

GIS Database Development and Exchange: Interaction Mechanisms and Motivations

Zorica Nedović-Budić, Jeffrey K. Pinto, and Lisa Warnecke

CONTENTS

ABSTRACT The idea of sharing geographic data both within and between organizations remains largely resisted despite the obvious benefits that can be derived from data sharing activities and federal initiatives that promote them. The research presented in this paper examines the various properties of data sharing activities, as well as related motivations cited by members of organizations as reasons for entering into cooperative relationships. The findings suggest that organizational members have a number of different reasons for engaging in data sharing relationships, with common missions/goals and saving of resources being the most frequently cited motivations. Financial resources are the more important reasons for external than for internal interactions. Further, both inter- and intraorganizational activities are guided with formalized mechanisms, the former predominantly in the form of legal contracts and agreements, and the latter predominantly in the form of policies and mutual rules and procedures. Informal interactions, however, continue to be significant in facilitating those interactions. We find that adoption of standards is still inadequate to enable ubiquitous data integration and exchange, but certainly appears stimulated by interorganizational engagements. In terms of contributions, the geographic data remain the main good exchanged. More involved interactions, such as coordinated database development and maintenance and joint applications and clearinghouses, are more likely to happen internally. The Internet, although still not prevalent, has started to facilitate communications and relationships with external partners. The implications of this research are considered as they relate to future efforts to induce wider sharing of geographic information system (GIS) data across organizational boundaries and to build spatial data infrastructures at all levels.

4.1 Introduction

In the United States, the Federal National Spatial Data Infrastructure (NSDI) initiative calls for the development of an "information highway" to connect the variety of spatial data producers and users, including government, private sector, and academic institutions (FGDC 1994). Initiatives on data sharing mechanisms, infrastructure, institutional arrangements, and standards, along with improvements in the enabling tools, are crucial for assisting data sharing practices and for building the NSDI (Nedović-Budić and Pinto 2001). Such efforts to achieve data coordination have been ongoing in the U.S. for many years through efforts to develop and distribute digital data. However, until recently there were few pronounced standards or mechanisms for coordinating spatial data resources between multiple levels of government, leading to rather isolated initiatives with respect to data integration. Nevertheless, some of the early data coordination projects started to lay an important foundation for future efforts in the area. The development of GBF/DIME and TIGER files, for instance, promoted data sharing among more than three hundred local planning agencies and managed to overcome bureaucratic inertia (Sperling 1995). Most recently, the initiative for development of the National Digital Geospatial Data Framework and Clearinghouse and Geospatial One-Stop web portal (FGDC 1994; 2003) and the National Map project of the U.S. Geological Survey (USGS 2002) are significant steps forward in inciting the comprehensive provision and exchange of reliable data among a variety of users, including all levels of government, private sector, and utilities. However, many of these initiatives lack implementation plans and tools, particularly with respect to the local and regional levels where a great majority of data production and use occurs. This situation leaves a satisfactory level of data integration across organizations and jurisdictional boundaries yet to be realized (Haithcoat et al. 2001).

Geographic information systems (GIS) technology enables data integration across organizations (Campbell and Masser 1991) and can stimulate interorganizational alliances (Kumar and Van Dissel 1996; Roche and Humeau 1999; Dedekorkut 2002). Infrastructures such as the NSDI would allow for unlimited sharing of spatial data, and thus prevent duplication of effort and redundancy in developing geographic databases. Interestingly, however, the rapid increase in organizations adopting GIS technology has highlighted the fact that between and within organizations, there has been a general inability and often unwillingness to share data and information across boundaries, with concomitantly low levels of coordination (Warnecke et al. 1998). The problems here are typically not of a technical nature, reflecting instead a variety of "human" reasons why information continues to be hoarded and organizations resist seemingly obvious benefits of sharing data (Greenwald 2000; Nedović-Budić and Pinto 2000; Feick and Hall 1999). The waste caused by duplication of effort, due largely to lack of information exchange among local, state, federal government and private sector organizations, remains a significant impediment to the more effective and efficient use of GIS throughout society and hinders the development and utilization of the technology's full potential (Frank 1992; Warnecke 1999). Other recent work has further illustrated the incredible complexity that underscores efforts at data coordination. Francis Harvey's (2001a) survey in Kentucky and summary update of progress toward the National Digital Geospatial Data Framework and Clearinghouse demonstrate the complexity of GIS actor networks in collaborative environments. These findings were reinforced by Tulloch and Fuld's (2001) summary of the FGDC's survey of over 800 county-level data producers. They found that the needs for and types of data created by their sample were widely diverse, suggesting that the viability of achieving a framework data set remains highly complex.

Although recent research has offered some insights into the structure and motivations underlying information sharing across interorganizational boundaries (Nedović-Budić and Pinto 2000), empirical research to date has been primarily qualitative, relying on limited sets of case studies or anecdotal evidence. Noticeably lacking have been any larger-scale empirical research studies to investigate the nature of data sharing activities, including their primary characteristics. Though many researchers discuss "data sharing," of paramount concern is the question as to whether there is a clear agreement on what data sharing actually implies. This paper reports on the results of a recent research project that investigated the data sharing phenomenon across multiple public agencies on a U.S. national level. The results of this research offer an inside look at the current nature of many data sharing arrangements, in terms of variables such as the reasons for data sharing, the extent and nature of interaction, standardization activities, and levels of participation and contribution. Further, the research findings offer evidence of a direct link between specific motivations and structural variations in the data sharing arrangements.

4.2 Motivations and Interaction Mechanisms—Concepts and Theory

Coordinated systems and databases promise to stimulate interorganizational cooperation and collaboration and are expected to result in the provision of a better information base for management and strategic decision-making. Based on case study research conducted among U.S. city and/or county governments and other associated organizations, Nedović-Budić and Pinto (2000) identified two factors that shape the processes involved in data sharing activities as well as their outcomes: 1) motivations for engaging in data sharing activities, and 2) structural characteristics of the interaction mechanisms implemented by the data sharing entities.

4.2.1 Motivations for Data Sharing

The underlying assumption behind data-sharing initiatives is that such interorganizational sharing processes provide a number of benefits to the involved organizations. Avoidance of unnecessary data redundancy and duplication of efforts appear to be the most important goals of data coordination initiatives. The expected benefits from sharing, however, go beyond efficiency (Nedović-Budić and Pinto 2001). Benefits that are typically identified in the literature as the primary drivers for data exchange include the following (Nedović-Budić and Pinto 2000):

1. *Cost savings*—as consortia of agencies or independent organizations share data, they no longer need to duplicate data gathering and archiving, which leads to savings in terms of personnel, space/facilities, data acquisition and maintenance costs;
2. *Improved data availability*—a data archive could contain a larger selection of records than would be held by any one organization, thereby offering a more comprehensive library of geographic information;
3. *Enhanced interorganizational relationships*—underlying data sharing is the larger issue of promoting greater cross-organizational communication. It is assumed that among organizations that communicate and share information, there is a stronger opportunity to develop new, joint service missions within their jurisdictions.

Although cost savings are often mentioned as the major reason for interorganizational engagements (Nedović-Budić and Pinto 1999a), expected monetary benefits are not the only motivators for the establishment of data sharing mechanisms. In the GIS literature, the following additional reasons are cited as motivating GIS related data exchange: organizational needs and capabilities (Calkins and Weatherbe 1995); power relationships; appeals to professionalism and common goals (Obermeyer 1995); and incentives, superordinate goals, accessibility, as well as resource scarcity (Pinto and Onsrud 1995). O'Toole and Montjoy (1984) summarize the various motivations into three categories of inducements: 1) authority, 2) common interest, and 3) exchange (receiving something in return). It is important to consider these various types of motivations for data exchange as they can be expected to uniquely shape the structure of data sharing agreements. However, a comprehensive, empirically based account of reasons for establishing exchange or coordination mechanisms and impacts of such varying motivations on the nature and success of GIS development and exchange efforts are currently lacking in the respective literature.

4.2.2 Structural Characteristics of Interaction Mechanisms

Geographic data sharing efforts involve redefinition of existing tasks and structures and the establishment of new ones (Azad and Wiggins 1995). The structure of an interorganizational relationship is established by specifying roles, obligations, rights, procedures, locations, information flow, data, analysis and computational methods used in the relationship (Kumar and Van Dissel 1996). There are numerous ways to structure interorganizational GIS and database activities and the various configurations in which GIS resources are developed and exchanged often depend on the given institutional, technical, and economic constraints (Dueker 1987) as well as the above-mentioned specific motivations.

It is important to note that structure is a multi-dimensional concept. Five aspects of structure have been identified based on case study results as being of particular importance for studying interaction mechanisms (Nedović-Budić and Pinto 1999a; 1999b; 2000). First, distinct forms of sharing mechanisms can be identified based on the extent of interaction involved. Extent of interaction can vary from simple awareness of or communication about GIS activities to different forms of data exchange and, in its most sophisticated form, to joint system development and/or maintenance and sharing of resources beyond data (including personnel, facilities, and equipment). Second, interaction mechanisms can be either formal or informal in nature. Third, different standardization activities can be involved, ranging from no acceptance of standards to adoption of private, local, regional, federal or international standards. Fourth, a specific relationship can be characterized by the participation status of the entities involved. Different forms of membership with varying levels of influence or mere subscription status can be distinguished. Finally, the structure of the sharing relationship can be characterized based on the contributions of the specific entities to the joint GIS database development or data exchange activities.

4.2.3 A Conceptual Model of Data Sharing Motivations and Interaction Mechanisms

Based on this review of relevant literature, a conceptual model of data sharing mechanisms and their underlying motivations was developed (Figure 4.1). It assumes that the reasons for establishing data sharing agreements range from saving resources, common mission/goals, and existing organizational dependencies to various forms of directives from higher levels, disaster or emergency response/management, or a sense of duty for an authority. Further, the model suggests that the specific reasons behind data exchange relationships influence the specific structural layout of the interaction mechanisms implemented by the sharing entities. All five structural dimensions of interaction mechanisms identified above (extent of interaction, nature of interaction, standardization activities, participation status of entities involved, and level of contribution to shared activities) were incorporated in the model. Motivations as well as the resulting structural layout of the data sharing agreements are expected to differ for data sharing projects depending on whether the exchange occurs internally (within organizations) or externally (between organizations).

4.3 Methodology

A national survey of city and/or county governments, regional entities, and other related public and private organizations that engage in data sharing activities was conducted to validate the relationships proposed by the conceptual model. A self-administered survey instrument was designed based on items derived from the model concepts. The questionnaire was pre-tested through interviews with GIS professionals and academics for readability and clarity. The survey instrument is accessible at the following

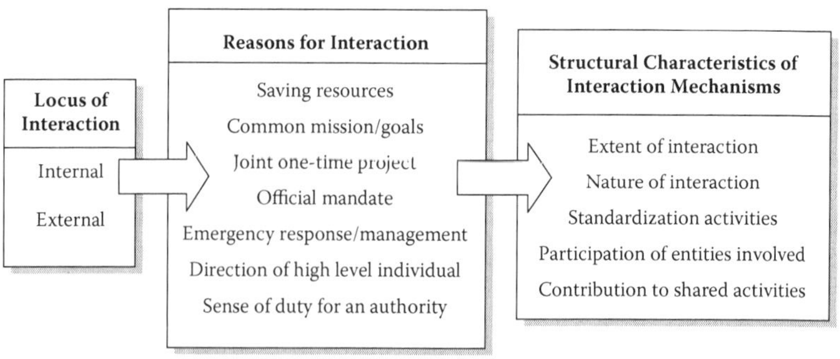

FIGURE 4.1
Conceptual model of relationships between motivations and interaction mechanisms.

address: http://www.urban.uiuc.edu/faculty/budic/W-NSF-2.html. Only the responses to Section 1 titled "Interorganizational Mechanisms and Motivations" are relevant for the discussions included in this paper. Factors related to context, relationship, implementation, and outcomes of interorganizational GIS were analyzed and presented in other works.

Questionnaires were distributed to a non-random, purposive sample of representatives from local governments involved in interorganizational data sharing relationships. The employees in GIS managerial role were targeted to fill out the questionnaire. Because the unit of analysis for this research was the individual organizational unit/entity from a sharing consortium or "cluster," the survey sampled 107 selected clusters of organizations using and sharing GIS across the United States. These clusters had been identified primarily based on the NSGIC/FGDC Framework Survey conducted in 1998 (FGDC 2002). Only sharing clusters that involved some kind of coordinating entity were included. Additional clusters known from the anecdotal GIS case study literature (e.g. conference proceedings and trade magazines) were also considered. Next, telephone interviews were conducted for screening purposes as only clusters that reported interorganizational data sharing activities were to be included in the final sample. This initial screening was also used to obtain contact details for the other sharing entities within the cluster. At least five units/entities per cluster were contacted to account for differences in the views of the members of the sharing relationship. In most states, two clusters were identified and sampled, although in more populous states three to six sites were sampled. In less populous states only one cluster was sampled. All states were represented in the sample.

Information about the relationships of the respondent organization with the other entities in the data sharing cluster was collected by asking the respondents to consider a maximum of eight entities *internal* to and eight entities *external* to their organization. Departments, divisions, or other units within an organization (for example, within city or county government) were considered as internal to that organization; functionally, financially, and administratively independent organizations were considered as external entities (e.g., utility companies, municipal governments, regional planning commission, and school district would be regarded external to a county government). For each of the listed entities, the respondents were asked to characterize their shared geographic information systems/geographic data (GIS/GD) activities in terms of extent, reasons, and the nature of the interaction, as well as standardization activities, participant status, and contribution to shared GIS/GD activities by selecting all applicable responses. A respondent's interactions were identified as having certain characteristics if at least one of the internal interactions or at least one of the external interactions was described in that specific way.

A total of 529 questionnaires were mailed out following the Dilman (1978) total design method for mail survey administration. The final sample yielded

245 responses, for a response rate of 46 percent. The sample was analyzed with respect to the locations of the organizations to see whether all U.S. regions were represented in the data. For this purpose, the sample pool was defined geographically to include four regions as defined by the U.S. Census Bureau. The sample used for the analysis had the following geographical distribution: south (79 responses or 32.2 percent), west (58 responses or 23.7 percent), northeast (42 responses or 17.1 percent), and midwest (66 responses or 26.9 percent). In terms of types of organization represented in the respondent pool, city and county government responses outnumbered all other types by a large margin. A demographic breakdown of the sample in terms of organizational type is shown in Table 4.1. The majority of responses came from representatives from city, county, or regional governments.

To get a sense of the character of the area in which the clusters of GIS users surveyed were situated, responses were also broken down by population size. Medium size counties in categories 50,001–250,000 and 250,000–1 million inhabitants represent the majority of the sample—about 80 percent. More specifically, cluster locations with a population of less than 50,000 account for 12.7 percent of the sample and those with 50,001–250,000 inhabitants for 34.7 percent; 44.5 percent of the clusters have a population size of 250,001–1 million and only 8.2 percent of the locations have a population of more than 1 million inhabitants.

Finally, the respondents contained in the sample can also be described by organizational function (Table 4.2). While a substantial portion of respondents identified themselves as "planning" (24.6 percent), when the categories of official and unofficial GIS/IT organizational representatives are compiled, they represent over 30 percent of the overall sample.

The characteristics of the respondent organizations in terms of geographical distribution, population size, and type were compared to those organizations that had not responded to the survey. No significant differences were found. For the purpose of this paper, the specific characteristics of the organizations were only used to describe the sample and were not taken into account for further analyses. Rather, the focus was to point to differences

TABLE 4.1

Distribution of Responses by Organizational Type

Organizational Type	Respondents (n = 245)
City	94 (38.4%)
Joint city/county	13 (5.3%)
County	72 (29.4%)
Utility	9 (3.7%)
Regional	38 (15.5%)
Special purpose organization	5 (2.0%)
Other, for profit	6 (2.4%)
Other, non-profit	8 (3.3%)

TABLE 4.2

Distribution of Responses by Organizational Function

Organizational Function	Respondents (n = 245)	Organizational Function	Respondents (n = 245)
Elected official	3 (1.2%)	Planning	60 (24.6%)
Chief administrative office	7 (2.9%)	Building inspection	1 (0.4%)
Information systems office	21 (8.6%)	Public works	25 (10.2%)
Official GIS within IT	33 (13.5%)	Utilities	12 (4.9%)
Official GIS outside IT	32 (13.1%)	Environment/ natural res.	11 (4.5%)
Unofficial GIS within IT	4 (1.6%)	Property mgmt/ real estate	2 (0.8%)
Unofficial GIS outside IT	10 (4.1%)	Health and human services	2 (0.8%)
Finance	2 (0.8%)	Public safety	4 (1.6%)
Taxation	11 (4.5%)	Other	4 (1.6%)

between internal and external sharing activities, and it was assumed that these differences would be found for all organizations, regardless of their type, function, location, or size of jurisdiction. The variability across those factors was found sufficient not to bias the results. Even the respondents from the same cluster represented very different organizational units with different contributions to and perceptions of the shared GIS. Also, many different clusters were included to offer enough variance in values and no significant number of respondents belonged to the same cluster. In a majority of cases there were only two respondents per cluster.

Descriptive analyses were conducted to investigate the occurrence of certain data-related activities and types of interactions within the sample as well as to obtain insights with respect to the motivations that underlie such joint GIS development and data exchange activities. While the general descriptive statistics were generated for the whole pool of the respondents, for a cluster to be included in further analysis a minimum of two members of the interorganizational sharing cluster had to respond to ensure that each cluster was represented by at least two entities. However, only a few clusters had more than two respondents. Therefore, a dataset composed of 228 responses was used for a preliminary analysis of differences between internal and external interactions to gain a better understanding of the specific nature of interorganizational versus intraorganizational interaction mechanisms. A series of t-tests was conducted to analyze whether significant differences between internal and external relationships existed. Finally, the relationship between the most common motivations and the structural aspects of interactions was analyzed using cross-tabulations and chi square statistics.

4.4 Characterizing the Interactions

The reasons for establishing shared GIS development or data exchange activities, as depicted in Figure 4.2, vary widely, though the most frequently cited reasons include common goals or mission (73.7 percent), desire to save resources (47.3 percent), and functional dependency (29.1 percent). These findings reinforce early cross-functional cooperation by demonstrating that superordinate or jointly held and compelling goals are the strongest antecedent motivating collaboration among organizational sub-units (Pinto, Pinto, and Prescott 1993). When there exists a fundamental data exchange "motivator" for all concerned parties in the form of shared goals, economies of scales through saving resources (Croswell 1991), or functional dependency in the form of pooled, sequential, or reciprocal interdependency as noted by Thompson (1967), there is a strong incentive to engage in cooperative data-sharing ventures.

Many of the interactions captured in this study do not reach beyond basic levels (Figure 4.3). However, a large majority of sharing activities (76.4 percent) have at least reached a stage where actual data exchange occurs. Interestingly, about half (54.4 percent) of these data exchange relationships are described as being free of charge and only about a third (33.7 percent) of the exchanges are mutual. More complex forms of interaction, such as joint database development and the sharing of resources beyond data (such as personnel, facilities, or equipment) are less common. However, the findings indicate that efforts in this direction are being undertaken and that a significant number of sharing activities include at least some coordination in terms of database development. Nevertheless, more active sharing efforts that aim at facilitating a direct access to data, such as the development of

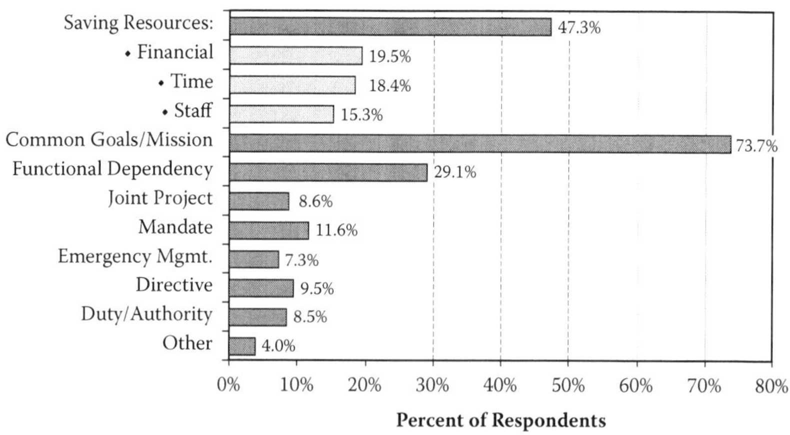

FIGURE 4.2
Reasons for interaction (n = 245).

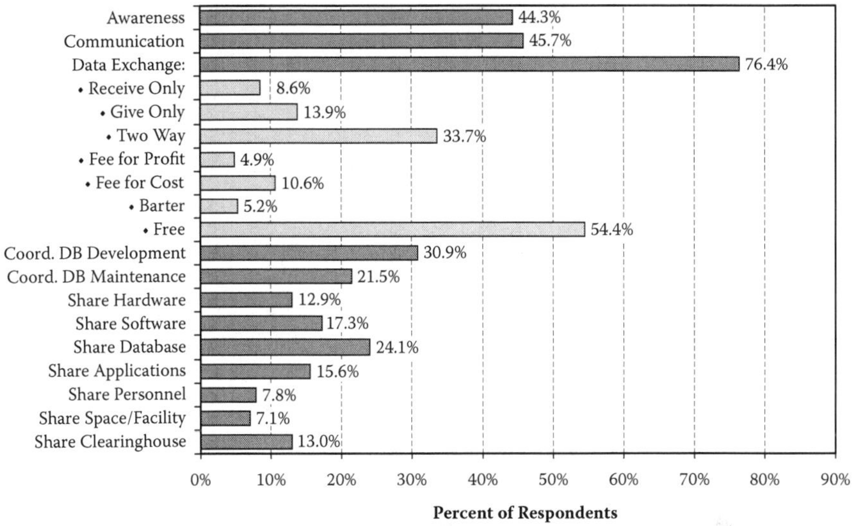

FIGURE 4.3
Extent of interaction (n = 245).

spatial data clearinghouses, appear to be still largely neglected, with only 13 percent of the interactions being focused on such initiatives. A recent national study of metropolitan planning organizations by Knaap and Nedović-Budić (2003) affirms the relatively low reliance on Internet and clearinghouses for data exchange or access, with less than one fifth of organizations reporting such practices. This particular finding confirms the trend toward individual GIS developments found in the FGDC survey (Tulloch and Fuld 2001) and a reliance on less efficient data access methods, which could potentially endanger investments in broader efforts to build spatial data infrastructures.

As we have previously noted, one area of possible misinterpretation has to do with divergent views as to what information sharing among interorganizational units actually connotes. That is, "data sharing" as a concept may be viewed benignly by various exchange partners; however, the practical mechanics of what data sharing actually requires tends to result in a potentially contentious process of proposal and counter-argument (Nedović-Budić and Pinto 2000). In the absence of clear guidelines as to the nature of data sharing interactions, there is a strong opportunity for political and power differentials to tilt the balance away from an equal exchange to one that may be beneficial only to one of two partners (Azad and Wiggins 1995; Pinto and Azad, 1994). Hence, another important finding from this research is to isolate the nature of the various forms of data sharing interactions that are most often practiced in these initiatives. The key mechanisms for coordination are shown in Figure 4.4.

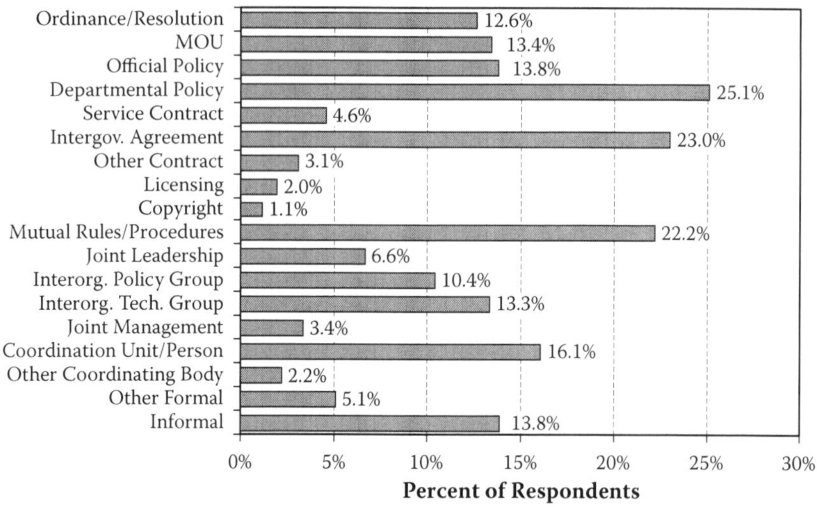

FIGURE 4.4
Nature of interaction (n = 245).

Only a small portion of the sharing relationships included in the analysis are informal (13.8 percent), i.e., ad hoc or based on personal contacts, needs, or availability. The most common formal types of agreement governing the involved interactions are "departmental policy" with 25.1 percent, followed by "intergovernmental agreement" (23 percent) and "mutually accepted rules and procedures" (22.2 percent). Joint leadership, coordinating units, and policy or technical groups are also present but do not govern many relationships. "Copyright" is the least frequently used governing mechanism with only about 1 percent of the relations being based on this type of arrangement. It appears that the sharing entities regulate the relationships rather than the content of their activities. The content, however, is often guarded by the use of disclaimers and distribution restrictions. Further, although a large majority of the interactions is governed by formal agreements rather than being driven by ad hoc needs, the border between informal and formal seems to be fuzzy and less formal ways of regulating the data sharing relationships, such as mutual rules and procedures, appear to be as important as very explicit mechanisms. This finding confirms the difficulty in distinguishing between formal and informal agreements as well as the relevance of rather informal sharing networks, particularly in smaller local governments, suggested by Harvey (2001a; 2001b).

The issue of standards selected is intriguing. Figure 4.5 gives the breakdowns of the types of standardization activities that underlie the sharing ventures studied. The most commonly used standards tend to be those developed locally (66.5 percent), rather than some "higher order" standard developed at the state, federal, or international level. In fact, outside of those agreed to by the sharing partners, most clusters are as likely to adopt no

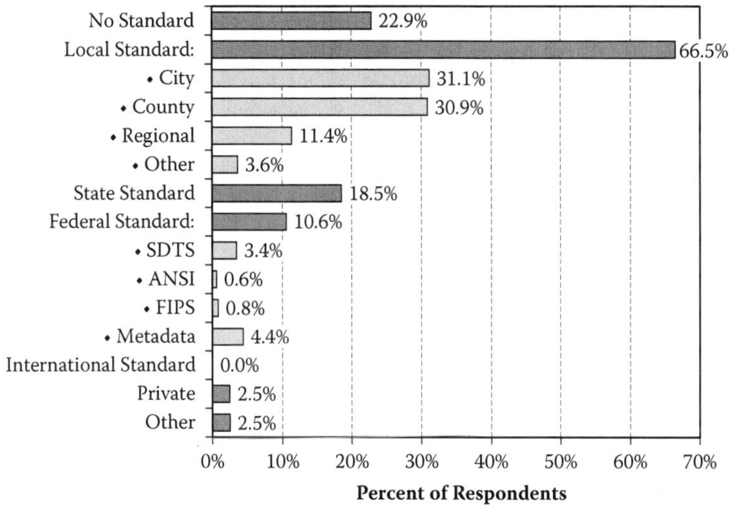

FIGURE 4.5
Type of standard employed (n = 245).

standard at all as they are to employ a state or federal standard. These findings offer further credence and support for the recent analyses of the FGDC survey findings by Harvey (2001a) and Tulloch and Fuld (2001), who note the tendency for coordination efforts to conflict with the complex reality of locally developed data and sharing practices. The study results suggest additional challenges that face proponents of NSDI and related initiatives, such as the National Map, given a tendency to resist such higher, national standards.

Another issue identified in the study is the nature/status of various participants in a data sharing arrangement. Do these clusters tend to be dominated by the main members, whose agenda could prevail over a number of less powerful members, or who could be operating within a more egalitarian participation process? Figure 4.6 shows findings with respect to participation status. From a demographic perspective, it is clear that the majority of the respondents to the survey (though not an absolute majority) see their participation as being represented best by the classification "lead member." Smaller proportions of respondents are non-members, subscribers, or having membership either with or without voting rights. The way in which respondents participate in the surveyed clusters is certainly expected to influence their position on various interorganizational issues and relationships.

An issue of significant interest has to do with the various contribution levels offered by members of these interorganizational networks; that is, to what degree and in what manner do most members contribute to the initiative? As Figure 4.7 demonstrates, the clear majority of participants identify their primary contribution as consisting of supplying geographic data for use by

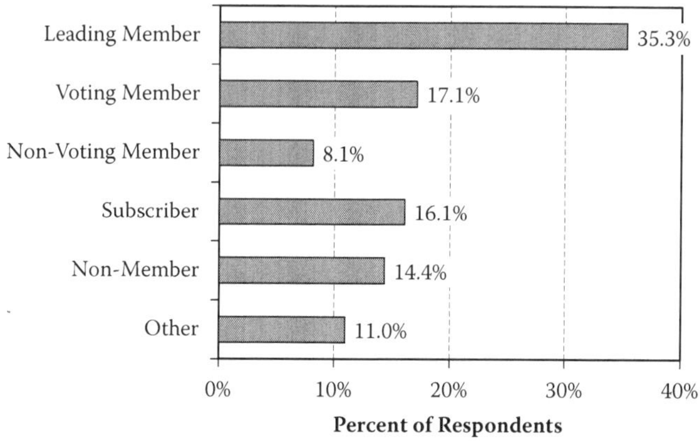

FIGURE 4.6
Participation status (n = 245).

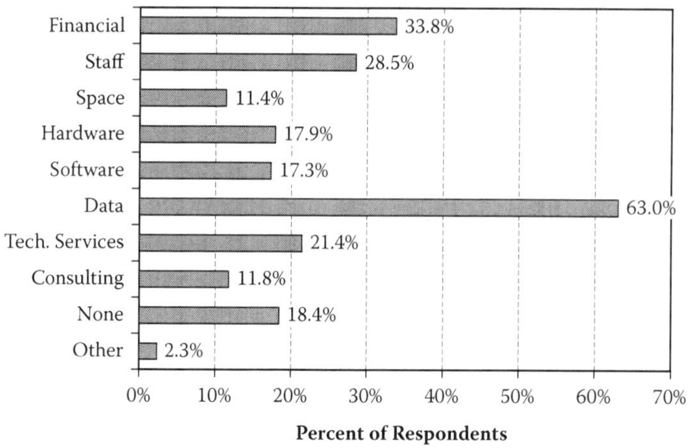

FIGURE 4.7
Contribution to shared GIS/GD activities (n = 245).

other organizations. This result confirms the relative importance of data-driven sharing relationships and a view of data as a public good (Masser and Campbell 1995). A significant percentage of the sharing partners also provide financial and/or staff support to the arrangement.

4.4.1 Intraorganizational versus Interorganizational Motivations and Interaction Mechanisms

When looking at issues of sharing from an intraorganizational versus inter-organizational perspective, i.e., by comparing relationships that are internal

TABLE 4.3

Reasons for Interaction

Reasons for Interaction	Percent of Respondents (n = 228)	
	Internal	External
Saving resources:	64.3	62.0
• Financial*	17.3	27.1
• Time	25.5	23.4
• Staff	22.6	21.9
Common goals/mission*	84.7	78.6
Functional dependency*	49.5	36.5
Joint project	12.2	16.1
Mandate	15.3	14.6
Emergency management*	17.9	10.9
Directive*	11.2	2.6
Duty/authority	12.8	16.7
Other	6.1	5.7

* Significant at the 0.05 level.

to the organization to those that cross organizational boundaries, some additional findings become evident. First, interesting differences in the reasons for sharing interactions were found. Saving financial resources is a much more frequent motivator for external interactions than for internal sharing agreements (Table 4.3). In contrast, saving time and staff are slightly more frequent motivators for internal relationships; however, the difference is not significant. Common goals, functional dependency, emergency management, and especially directives from higher-level organizations are more frequent reasons underlying internal than external relationships. Further, sense of duty for an authority is a more common motivator in external relationships, although this difference is not significant. Only slight differences are found for official mandate and joint one-time projects, with official mandate being more common in the intraorganizational context and joint projects being more frequently mentioned with external relationships.

Regarding the extent of interactions, coordinated database development and more complex forms of interaction beyond just the data (including software, hardware, applications, space, and staffing) are more evident with internal relationships (Table 4.4). Also, the use of a joint clearinghouse is more common within organizational boundaries, although the difference was not significant. Not surprisingly, both "fee for profit" and "fee for cost" arrangements are more often mentioned with external interactions. Free exchange relationships are slightly more common in the context of internal sharing. Further, one-way interactions that involve giving data are more prevalent with internal relationships whereas receiving data is more common if the exchange occurs with an external agency. Interestingly, city standards appear to play a more significant role in internal relationships while county

TABLE 4.4

Extent of Interaction and Standardization Activities for Internal versus External Interactions

Extent of Interaction	Percent of Responses (n = 228) Internal	External	Standardization Activities	Percent of Responses (n = 228) Internal	External
Awareness	56.9	58.6	No standard*	24.0	34.4
Communication	57.9	63.4	Local standard†:	70.4	64.1
Data exchange:	86.2	90.6	• City	32.7	29.2
• Receive only*	3.1	7.9	• County	29.1	33.9
• Give only*	25.6	12.6	• Regional*	10.2	21.4
• Two way	46.2	51.8	• Other	5.6	3.6
• Fee for profit*	0	2.6	State standard*	16.8	25.0
• Fee for cost*	5.1	25.7	Federal standard†:	9.7	14.6
• Barter	0	0	• SDTS	3.6	3.6
• Free	67.2	63.4	• ANSI	0.5	0.5
Coord. DB Development*	58.5	32.5	• FIPS	1.0	1.0
Coord. DB Maintenance*	43.1	22.5	• Metadata	3.6	4.2
Share hardware*	30.3	7.9	International standard	0	0
Share software*	37.9	9.9	Private	3.1	2.1
Share database*	46.7	20.4	Other	4.1	3.1
Share applications*	31.8	11.0			
Share personnel*	21.0	6.3			
Share space/facility*	15.9	2.1			
Share clearinghouse	18.5	15.7			

* Significant at the 0.05 level.
† Significant at the 0.1 level.

and regional standards are more frequently employed when the interaction occurs with external organizations or agencies. One has to, of course, take into account that a large portion of the organizations surveyed are city governments. Importantly, state and federal standards are more often used in conjunction with external sharing activities. Thus, it appears that the sharing entities recognize a greater need for such standards when engaging in interorganizational sharing.

The nature of interactions often varies widely depending upon whether the sharing arrangement is predominantly based on internal or on external partners (Table 4.5). Interestingly, for internal sharing relationships, the mechanisms tend to be departmental policies or official policies. On the other hand, for external relationships, we find a higher frequency of intergovernmental agreements and memos of understanding (MOUs), as well as service and other contracts. Further, it appears that coordination units and joint leadership and management are more common ways of coordinating intraorganizational sharing activities. Policy groups are more frequently mentioned in connection with internal sharing activities. The same is true for technical groups; however, the difference was not significant. Surprisingly, informal exchanges can be more frequently found in the interorganizational context; however, the difference was not significant.

In terms of participation status, "leading member" occurs more frequently in conjunction with internal interactions while the status "non-member" is more common for entities that are external (Table 4.5). The findings for contributions to the shared GIS activities confirm the results presented for the extent to which the different entities engage in sharing relationships. Sharing beyond simple data exchange, i.e., staff, space, hardware, and software contributions, is more common when the interaction is internal. Contributions in the form of financial resources and technical services can also be more frequently found in the internal context, yet the difference was not significant. Consulting is more prevalent in external activities. Again, the results illustrate that exchanges that only involve receiving data and do not involve active contributions from an entity occur somewhat more often in the interorganizational context.

4.4.2 Relating Motivations and Structures

Since many significant differences were found between internal and external interactions, the relationships between motivations and structural aspects of sharing were tested separately for intraorganizational and interorganizational relationships. The implications of the presence versus non-presence of the most common motivations were explored using cross-tabulations and chi-square statistics. A relationship was identified as being positive when the interactions that were characterized by that specific motivation were more frequently associated with the presence of a certain structural characteristic than those for which the given motivation was not present.

Only the three most common motivations—1) saving resources, 2) common goals/mission, and 3) functional dependency—were considered for this analysis because the number of responses for some of the less frequent reasons were too low to conduct cross-tabulations. Also, only significant relationships (based on chi-square tests) are indicated. The interrelationships between the existence of certain forms of contributions, standardization, and participation and the nature as well as the extent of interaction were investigated using the same methodology. The following text highlights some of the most important relationships; tables that display all significant relationships can be found in Appendix A.

Respondents motivated to participate in interorganizational relationships in order to save resources are more likely to establish closer interactions with other entities than respondents not motivated by saving resources. These interactions include communication; data sharing; joint database

TABLE 4.5

Nature of Interaction, Participation Status, and Contributions by Internal vs. External Interactions

	Percent of Responses (n = 228)			Percent of Responses (n = 228)	
Nature of Interaction	**Internal**	**External**	**Participation Status**	**Internal**	**External**
Ordinance/resolution	17.3	16.6	Leading member*	55.1	38.0
MOU*	7.1	29.5	Voting member	19.9	22.4
Official policy*	16.3	8.8	Non-voting member	13.8	13.5
Dept. policy*	31.6	22.3	Subscriber	15.8	21.9
Service contract*	2.0	8.3	Non-member*	11.2	30.7
Intergov. agreement*	18.4	40.9	Other	12.8	14.1
Other contract*	**0.5**	**5.2**	**Contributions**	**Internal**	**External**
Licensing	1.5	2.1			
Copyright	1.0	1.0	Financial	52.0	46.1
Mutual rules/procedures	31.6	30.6	Staff*	52.6	34.2
Joint leadership†	11.7	6.7	Space*	27.0	13.0
Interorg. policy group†	12.2	7.8	Hardware*	37.8	22.3
Interorg. tech. group	13.8	13.5	Software*	35.7	20.2
Joint management†	7.1	4.1	Data	72.4	73.6
Coordination unit/person*	21.9	13.0	Tech. services†	37.8	33.7
Other coord. body	1.5	3.6	Consulting*	14.8	22.8
Other formal governance*	4.6	9.3	None	26.5	32.1
Informal	13.8	17.6	Other	3.1	3.1

* Significant at the 0.05 level.
† Significant at the 0.1 level.

development; sharing of hardware/software, space, and personnel; and, in the case of external entities, joint applications and clearinghouses. Saving of resources is also the strongest driver for the use of standards, internally in particular. This motivator coincides with the use of mutual rules/procedures and joint leadership as the most prominent internal interaction mechanisms, with generally less emphasis on formalization. More formal mechanisms, such as intergovernmental agreements, are more common for external interactions, which still rely considerably on informal relationships. Respondents who express that saving resources is their motivation for GIS interactions contribute in various ways, but more extensively in internal settings.

Common goals/mission tends to stimulate two-way exchanges and establishment of clearinghouses, while functional dependency internally appears to allow for sharing of space and personnel, and externally for sharing and joint database developments. Common goals/mission as the motivator is associated with state standards in both internal and external contexts. In terms of mechanisms, functional dependency tends to stimulate the use of policy and technical groups for interaction mechanisms. Among the varying contributions by internal and external entities, the contributions that stand out are financial in the case of interactions based on goals/mission as motivation and staff in the case of interactions based on functional dependency. Data remains the main contribution item in all relationships that were included in the analysis.

In exploring the relationship among various structural factors, we find that respondents in the lead member role report extensive relationships, including establishment of clearinghouses, particularly in relationships with external entities. In those external settings the emphasis is more on two-way exchanges. Standardization efforts also relate to closer relationships, which for external entities tend to include sharing of hardware and software. The two main types of contributions— financial and data—tend to associate with more intense relationships, the former in external settings, and the latter in internal settings. The financial contribution also drives varied approaches to the cost attached to data exchanges—free for internal and fee-based for external exchanges. Data contributions internally are associated with joint applications and coordinated database maintenance; externally they coincide with coordinated database development and establishment of clearinghouses.

Finally, the nature of interactions indicated by the respondents in the role of lead members is more policy based in internal settings and more contract based in external settings. A very similar pattern is found with respect to both financial and data contributions. The presence of local standards is associated with formalized mechanisms both internally and externally, including memoranda of understanding and intergovernmental agreements. However, the persisting characteristic of the relationships applied in interactions with external entities is the mix of formal and informal mechanisms.

4.4.3 Summary

Respondents to the survey come from organizations with varying participation status in GIS sharing interactions. About a third of the respondents indicated their organization was a lead member, while the remainder includes other voting or non-voting members and subscribers. The respondents cite common goals/mission, saving of resources, and functional interdependences as the key motivation factors for their engagement in GIS and database sharing activities. Among various organizational resources, financial resources seem to be more relevant than staff and time resources for external relationships, while directives and emergency management are more stimulating internally.

Overall, data is the main currency in GIS sharing activities. It is mostly exchanged for free and two way, particularly in internal settings. Externally, the fee-based exchange is more present. Coordinated database development or maintenance is pursued, but somewhat sporadically, especially when external partners are involved. Similarly, the efforts to establish clearinghouses are irregular, although they are more likely to happen internally. Externally, the clearinghouses are more likely to be set up when the motivation for the relationship is to save resources than in case of other motivations. However, in a recent study Knapp and Nedović-Budić (2003) suggest that although the reliance on the Internet for data access or exchange is becoming more frequent, it is still not a dominant practice among external partnering organizations.

To conduct GIS sharing activities, organizations use various but mostly formal interaction mechanisms, including: policies, intergovernmental agreements, mutual rules and procedures, memoranda of understanding, ordinances and resolutions, policy and/or technical groups, and joint coordination unit/person. Internal relationships are more based on policy, groups, and joint management, leadership, or coordination point; external relationships are more based on agreements, memoranda of understanding, and contracts. Interestingly, external relationships are more formalized, but also rely on informal interactions. In general, the entities involved in GIS and database sharing activities tend to regulate their relationships more than the content of data that is exchanged.

Standardization activities are characterized primarily by the use of locally developed standards. The higher use of county, regional, state, and federal standards in the external context is a promising indication that standards are recognized as necessary in facilitating cross-organizational exchange. Analysis of the motivations and structures points to saving resources as driving the use of standards internally (i.e., within an organization) and common goals/mission as stimulating the use of standards in coordinating or exchanging data with external organizational partners. Reliance on standards is also more present in formalized contract-based interactions and among organizations that contribute financially.

Overall, the saving of resources as the motivator and data as the form of contribution have strong presence across interorganizational and intraorganizational settings, but they are a more typical factor in internal organizational bonding. Common goals/mission is a stronger motivator of external relationships, most often realized through data exchanges and establishment of joint clearinghouses. In those interorganizational settings, financial contributions tend to create tighter ties between external organizations, but also prompt the fee-based data exchanges.

4.5 Conclusions

While there is general agreement that data sharing does represent a positive step forward in advancing the goals of the NSDI as well as in promoting greater joint cooperation among distinct organizational units, one of the frustrations to date has been the general lack of empirical evidence to support the more widely held assumptions regarding the characteristics of data sharing activities. Previous research that has employed case studies or anecdotal evidence, though offering some valuable preliminary evidence of the nature of cooperative data sharing ventures, could not provide a comprehensive understanding of these interorganizational relationships. This research, employing a large-scale mail survey methodology based on previous qualitative findings, represents an important contribution toward improving our understanding of GIS data sharing.

One of the prevailing questions in studying interorganizational data sharing as it applies to geographic information has to do with gaining a better understanding of the nature of such sharing arrangements. While the concept of data sharing is becoming more and more accepted, our research and experience had led us to conclude that the manner in which such concepts as "sharing" are interpreted tended to vary widely, depending upon the individual and organizational unit. For some, cooperation seems to be defined simply as "non-aggressive" coexistence while others hold a more positive, resource linkage-based view. This research sought to directly address a number of the principal questions underlying GIS data sharing interactions among organizations in terms of why they shared data, the extent of sharing activities, the nature of the sharing relationship, the standards employed, the participation status of the units/entities involved, and the contributions made. As a result, this research offers a unique glimpse inside the characteristics of and motivations surrounding organizational units willing to enter into data sharing activities across organizational boundaries.

The results presented in this paper indicate that:

- joint mission/goals, saving resources, and existing functional dependencies are the most common motivations that drive data sharing activities;
- exchange relationships are most often restricted to simple data sharing and are frequently provided or received free of charge;
- building of data clearinghouses continues to be somewhat sporadic, regardless of the increased use and role of the Internet in accessing and exchanging information;
- if standards are employed, they are mostly locally agreed upon rather than based on national or international standards;
- organizations tend to regulate their relationship more than the contents of their exchanges; ordinances/resolutions, departmental policies, MOUs, and intergovernmental agreements constitute the most commonly used formal mechanisms; other formal sharing agreements such as service contracts, licensing, and copyright agreements are not very frequently used;
- informal interactions still play a significant role in enabling data sharing.

All these characteristics of present sharing interactions indicate that a broader data sharing vision and related practices have not yet been adopted or have at least not been successfully implemented by the majority of public agencies in the United States. Further, significant differences were identified between interactions that occur in an intraorganizational context as opposed to sharing relationships with external entities. The findings indicate that:

- more active and extensive sharing is more likely to take place within organizational boundaries;
- this emphasis on internal sharing has led to a considerable lack of recognition for federal and international standards; standardization activities are more pronounced as part of external interactions and relationships;
- saving of financial resources is the most stimulating factor for enticing relationships with external organizations;
- exchange with external entities tends to be more fee-based than exchange with internal entities;
- internal interaction mechanisms are more policy-based, while external mechanisms are more likely to include a legal component; external interactions are more formalized, but still substantially taking advantage of informal relationships.

Yet, the study results also illustrate that motivations have a significant influence on the structural characteristics of the sharing activities. They also point to the relationships between particular structural elements. For instance:

- the presence of functional dependency as a motivating factor for internal relationships is found to be associated with limited one-way provision of data rather than comprehensive sharing activities;
- two-way exchanges are more present when joint mission/goals is a motivating factor;
- organizations contribute financially more often if joint mission/goals motivate their interorganizational relationships;
- extensive interactions are associated with more two-way exchanges, increased reliance on standards, and more substantial contributions to the joint geographic information activities.

The findings with respect to relationships between motivations and sharing mechanisms confirm that differences exist between the internal and external sharing context. Although additional research is needed to investigate interdependencies among the various motivations and the strengths of the particular relationships, the current findings provide an encouraging initial evidence for the assumption that instilling certain motivations into the data sharing communities could lead to more extensive sharing with a broader focus.

Given the nature of the data sharing agreements discovered as a result of our study, future research needs to continue to refine and sharpen definitions of such cooperative ventures. Due to the pressures to cut costs as well as to increase organizational efficiency while enhancing customer response and external effectiveness, the need to share geographic data across organizational boundaries is likely to increase rapidly. This research offers some valuable insights in terms of how such agreements currently work and, by implication, can suggest ways in which these agreements can be enhanced to derive the maximum advantage out of data sharing, both in terms of its impact on customer satisfaction as well as the promotion of greater cross-organizational cooperation and positive working relationships. Specifically, the importance of common goals/mission as a motivator for sharing activities and the more extensive interactions with internal partners suggest that successful sharing implies more than just the resolution of data-related issues. Also, the dominance of local standards in internal sharing arrangements suggests that much groundwork beyond the introduction of standards has yet to be done to establish broader sharing and commitment among the users of geographic data. Similarly, the continued low use of clearinghouses, especially with external partners, implies that data exchange currently occurs largely within local "islands of sharing" rather than on a regional or even national level. With no dramatic change occurring since the administration of this survey, Knaap and Nedović-Budić (2003) suggest a small positive trend in the reliance on the Internet for data access or exchange with external partners.

It is argued here that an extensive communication of a more comprehensive vision of data sharing is necessary and should make common goals/mission more explicit as it seems to serve as a very strong motivating factor. What is also necessary in addition to the vision is an implementation strategy and plan that incorporates these goals. It appears that understanding the nature and characteristics of the institutions involved in data sharing activities is crucial to understanding and fostering sharing relationships, especially in an interorganizational context. Further, the differences in internal versus external sharing relationships have to be considered and addressed in the design, implementation, and communication of national data sharing initiatives. Finally, future research has to concentrate on linking these situational and structural aspects to variables that measure the success of and/or satisfaction with such shared activities to be able to promote specific sharing arrangements that are most beneficial for certain organizational contexts.

Acknowledgment

This material is based upon work supported by the National Science Foundation under Grant No. IIS-9872015. Any opinions, findings, and conclusions or recommendations expressed in this material are those of the authors and do not necessarily reflect the views of the National Science Foundation.

References

Azad, B., and L. L. Wiggins, 1995, Dynamics of Inter-Organizational Geographic Data Sharing: A Conceptual Framework for Research. *Sharing Geographic Information*, eds. H. J. Onsrud and G. Rushton, 22–43. New Brunswick, NJ: Center for Urban Policy Research.

Calkins, H. W., and R. Weatherbe, 1995, Taxonomy of Spatial Data Sharing. *Sharing Geographic Information*, eds. H. J. Onsrud and G. Rushton, 65–75. New Brunswick, NJ: Center for Urban Policy Research.

Campbell, H. J., and I. Masser, 1991, The Impact of GIS on Local Government in Great Britain. *Proceedings of the Association for Geographic Information Conference*, London: AGI.

Croswell, P. L., 1991, Obstacles to GIS Implementation and Guidelines to Increase the Opportunities for Success. *Journal of the Urban and Regional Information Systems Association* 3, no. 1: 43–56.

Dedekorkut, A., 2002, What Makes Collaborations Work? An Empirical Assessment of Determinants of Success in Interorganizational Collaboration. Paper presented at the 44th Annual Conference of ACSP, Baltimore, Maryland, November 21–24, 2002.

Dilman, D. A., 1978, *Mail and Telephone Surveys: The Total Design Method.* New York: Wiley-Interscience.

Dueker, K. J., 1987, Multipurpose Land Information Systems: Technical, Economic, and Institutional Issues. *Photogrammetric Engineering and Remote Sensing* 53, no. 10: 1361–1365.

Federal Geographic Data Committee (FGDC), 1994, Development of a National Digital Geospatial Data Framework. *Status Report from the Framework Working Group.*

Federal Geographic Data Committee (FGDC), 2002, NSGIC/FGDC Framework Survey; http://www.fgdc.gov/framework/framework.html

Federal Geographic Data Committee (FGDC), 2003, Geospatial One-Stop; http://www.geo-one-stop.gov/

Feick, R. D., and G. B. Hall. 1999, Consensus-Building in a Multi-Participant Spatial Decision Support System. *Journal of Urban and Regional Information Systems Association* 11(2): 17–23.

Frank, A. U. 1992, Acquiring a Digital Base Map—A Theoretical Investigation Into a Form of Sharing Data. *Journal of Urban and Regional Information Systems Association* 4(1): 10–23.

Greenwald, M. J. 2000, Beyond City Limits: The Multi-Jurisdictional Applications of GIS. *Journal of Urban and Regional Information Systems Association* 12(1): 31–43.

Haithcoat, T., L. Warnecke, and Z. Nedović-Budić, 2001, Geographic Information Technology in Local Government: Experience and Issues. *The Municipal Year Book 2001.* Washington, DC: International City/County Management Association.

Harvey, F., 2001a, Constructing GIS: Actor Networks of Collaboration. *URISA Journal* 13, no. 1: 29–37.

Harvey, F., 2001b, NSDI from the Trenches—Local Government Perspective. *Geospatial Solutions* 11, no. 5: 2–4.

Knaap, G. and Z. Nedović-Budić, 2003, Assessment of Regional GIS Capacity for Transportation and Land Use Planning. Project report (draft) to the Lincoln Institute for Land Policy, U.S. HUD, and U.S. DOT. College Park, MD: University of Maryland, National Center for Smart Growth and Champaign, IL: University of Illinois at Urbana-Champaign, Department of Urban and Regional Planning. www.urban.uiuc.edu/faculty/budic/W-metroGIS.htm

Kumar, K., and H. G. van Dissel, 1996, Sustainable Collaboration: Managing Conflict and Cooperation in Interorganizational Systems. *MIS Quarterly* 20, no. 3: 279–300.

Masser, I., and H. J. Campbell, 1995, Information Sharing: The Effects of GIS on British Local Government. *Sharing Geographic Information,* eds. H. J. Onsrud and G. Rushton, 230–249. New Brunswick, NJ: Center for Urban Policy Research.

Nedović-Budić, Z., and J. K. Pinto, 2001, Organizational (Soft) GIS Interoperability: Lessons from the U.S. *International Journal of Applied Earth Observation and Geoinformation* 3, no. 3: 290–298.

Nedović-Budić, Z., and J. K. Pinto, 2000, Information Sharing in an Interorganizational GIS Environment. *Environment and Planning B: Planning and Design* 27, 455–474.

Nedović-Budić, Z., and J. K. Pinto, 1999a, Understanding Interorganizational GIS Activities: A Conceptual Framework. *Journal of the Urban and Regional Information System Association (URISA)* 11, no. 1: 53–64.

Nedović-Budić, Z., and J. K. Pinto, 1999b, Interorganizational GIS: Issues and Prospects. *Annals of Regional Science* 33, 183–195.

Obermeyer, N. J., 1995, Reducing Inter-Organizational Conflict to Facilitate Sharing Geographic Information. *Sharing Geographic Information,* eds. H. J. Onsrud and G. Rushton, 138–148. New Brunswick, NJ: Center for Urban Policy Research.

O'Toole, L. J., Jr., and R. S. Montjoy, 1984, Interorganizational Policy Implementation: A Theoretical Perspective. *Public Administration Review,* Nov/Dec: 491–505.

Pinto, J. K., and B. Azad, 1994, The Role of Organizational Politics in GIS Implementation. *Journal of the Urban and Regional Information Systems Association* 6, no. 2: 35–61.

Pinto, J. K., and H. J. Onsrud, 1995, Sharing Geographic Information across Organizational Boundaries: A Research Framework. *Sharing Geographic Information,* eds. H. J. Onsrud and G. Rushton, 44–64. New Brunswick, NJ: Center for Urban Policy Research.

Pinto, M. B., J. K. Pinto, and J. E. Prescott, 1993, Antecedents and Consequences of Project Team Cross-functional Cooperation. *Management Science* 39, 1281–1297.

Roche, S., and J. B. Humeau, 1999, GIS Development and Planning Collaboration: A Few Examples from France. *Journal of the Urban and Regional Information Systems Association (URISA)* 11, no. 1: 5–14.

Sperling, J., 1995, Development and Maintenance of the TIGER Database: Experiences in Spatial Data Sharing at the U.S. Bureau of the Census. *Sharing Geographic Information,* eds. H. J. Onsrud, and G. Rushton, 377–96. New Brunswick, NJ: Center for Urban Policy Research.

Thompson, J., 1967, *Organizations in Action.* New York: McGraw–Hill.

Tulloch, D., and J. Fuld, 2001, Exploring County-level Production of Framework Data: Analysis of the National Framework Data Survey. *Journal of the Urban and Regional Information Systems Association* 13, no. 2: 11–21.

United States Geological Survey (USGS), 2002, The National Map, http://www.nationalmap.usgs.gov/index.html

Warnecke, L., 1999, Geographic Information Technology Institutionalization in the Nation's States and Localities. *Photogrammetric Engineering & Remote Sensing* 65 (11), 1257–1268.

Warnecke, L., J. Beattie, K. Cheryl, W. Lyday, and S. French, 1998, *Geographic Information Technology in Cities and Counties: A Nationwide Assessment.* Washington, DC: American Forests.

Appendix A: Relating Motivations and Structures

TABLE A1

Relationship between Motivations and Extent of Interaction/Standardization Activities

	Internal			External		
	Saving resources	**Common goals/ mission**	**Functional dependency**	**Saving resources**	**Common goals/ mission**	**Functional dependency**
Extent of interaction						
Awareness						
Communication	P*	P*		P*	P*	P*
Data exchange:	P*			P*	P*	P*
• Receive only						
• Give only			P*			
• Two way		P†		P*		
• Fee for profit						
• Fee for cost						
• Barter						
• Free						
Coord. DB development	P*	P*		P*	P*	P*
Coord. DB maintenance	P*	P*		P*		
Share hardware	P†	P*				P*
Share software	P*	P*				
Share database	P*			P†		P*
Share applications				P*		
Share personnel			P*			
Share space/facility	P†		P*			
Share clearinghouse				P*	P*	

(Continued)

TABLE A1 (Continued)

Relationship between Motivations and Extent of Interaction/Standardization Activities

	Internal			External		
	Saving resources	**Common goals/ mission**	**Functional dependency**	**Saving resources**	**Common goals/ mission**	**Functional dependency**
Standardization activities						
No standard	N*					
Local standard:	P*	P†	P*	P*	P*	P*
• City	P*					
• County	P†			P†		
• Regional	P†	P*		P*		P†
• Other						
State standard		P*			P†	
Federal standard:	P*					
• SDTS						
• ANSI						
• FIPS						
• Metadata						
International standard						
Private						

P = positive association.
N = negative association.
* Significant at the 0.05 level.
† Significant at the 0.1 level.

TABLE A2

Relationship between Motivations and Nature of Interaction, Participation, and Contributions

	Internal			External		
	Saving resources	**Common goals/ mission**	**Functional dependency**	**Saving resources**	**Common goals/ mission**	**Functional dependency**
Nature of interaction						
Ordinance/resolution		P*				P*
MOU						
Official policy						
Dept. policy		P†				
Service contract						
Intergov. agreement					P*	P*
Other contract						
Licensing					N*	
Copyright						
Mutual rules/procedures	P*	P†			P†	P*
Joint leadership	P*	P†	P*			
Interorg. policy group		P*	P†			P*
Interorg. tech. group		P*	P*			P†
Joint management						
Coordination unit/ person						
Other coord. body						
Other formal governance			N†			
Informal					P*	P†

(Continued)

TABLE A2 (Continued)

Relationship between Motivations and Nature of Interaction, Participation, and Contributions

	Internal			External		
	Saving resources	**Common goals/ mission**	**Functional dependency**	**Saving resources**	**Common goals/ mission**	**Functional dependency**
Participation status						
Leading member		P†	P*	P*		P*
Voting member						
Non-voting member						
Subscriber	P†					
Non-member						
Contributions						
Financial	P*	P*	P*	P*	P*	
Staff	P*	P*	P*	P*		P*
Space	P*		P*	P*		
Hardware	P*	P*	P*			
Software	P*	P*	P*	P*	P*	
Data	P*	P*	P*	P*	P*	P†
Tech. services	P*	P*	P*	P*	P†	P*
Consulting				P*		
None						
Other	N*					

P = positive association.
N = negative association.
* Significant at the 0.05 level.
† Significant at the 0.1 level.

TABLE A3

Relationship between Participation (Leading Member), Standardization (Local), and Financial and Data Contributions and Extent of Interaction

	Internal				External			
	Leading member	**Local standard**	**Financial contrib.**	**Data contrib.**	**Leading member**	**Local standard**	**Financial contrib.**	**Data contrib.**
Extent of interaction								
Awareness	N*							
Communication			P*		P*		P*	P†
Data exchange:			P†		P†	P*		P†
• Give only					N†			
• Two way				P*	P†	P*		
• Fee for cost							P†	
• Free			P*	P*				
Coord. DB development		P*	P*		P*	P*	P*	P*
Coord. DB maintenance	P*	P*	P*	P*	P*	P*	P*	
Share hardware						P†		
Share software	P*	P*	P*			P*		
Share database	P*	P*	P*		P*	P*	P*	P*
Share applications	P†	P*	P*	P†	P*	P*	P*	
Share personnel			P*		P†		P*	
Share space/facility			P†		P*			
Share clearinghouse	P†				P*	P*	P*	P*

P = positive association.
N = negative association.
* Significant at the 0.05 level.
† Significant at the 0.1 level.

TABLE A4

Relationship between Participation (Leading Member), Standardization (Local), and Financial and Data Contributions and Nature of Interaction

	Internal				External			
	Leading member	Local standard	Financial contrib.	Data contrib.	Leading member	Local standard	Financial contrib.	Data contrib.
Nature of interaction								
Ordinance/resolution					P*	P*	P*	
MOU		P†				P*	P†	
Dept. policy	P*		P*	P†				
Service contract		N†						
Intergov. agreement		P*				P*		P†
Other contract					N†			
Mutual rules/ procedures				P†	P*			P*
Joint leadership		P*	P†				P†	
Interorg. policy group	P*			P*	P*		P*	
Interorg. tech. group					P†			
Coordination unit/ person								P†
Informal	N†							P*

P = positive association.
N = negative association.
* Significant at the 0.05 level.
† Significant at the 0.1 level.

Section 2

Data and Technology

5

SDI Reality in Uganda: Coordinating between Redundancy and Efficiency

Walter T. de Vries and Kate T. Lance

CONTENTS

5.1 Introduction

Coordination of spatial data infrastructures (SDIs) is a balancing act between technological efficiency objectives and public sector inefficiency realities. SDI technological advocates typically promote that SDI coordination should aim for national, seamless, standardized, and nonredundant data sets. This would increase operational efficiencies when organizations need to share spatial data and would avoid the additional work and transaction costs of data duplication (Rasmussen 1993; Buogo and Chevallier 1995; Fonseca et al. 2000; Astle et al. 2006; Baker and Chandler 2008). Yet existing public sector contexts can have inherent drivers toward maintaining inefficiency and redundancy. These include the need to construct backup mechanisms in case of organizational uncertainties such as the lack of public sector resources, and the need to be flexible in case of political uncertainties, such as government failure and political transitions (Landau 1969; Miranda and Lerner 1995; Ting 2003).

Consequently, SDI coordination is caught in an implementation quandary of establishing efficient and nonredundant spatial data processes versus inherently inefficient and possibly redundant public sector needs. This invariably stymies SDI development. To advance the knowledge of SDI development, we need a greater understanding of the public sector context in which SDI coordination is embedded.

Understanding the public sector context is particularly critical for SDI coordination in developing countries, given that the public sector system in developing countries is more prone to inefficiency drivers than it is in developed countries. The drivers of the public sector in developed countries include being flexible to heterogeneous citizens' requirements, being adaptable in case of political transitions, and being plural in executing multiple public tasks and responsibilities simultaneously (Vigoda 2002; Bekkers 2009). However, in developing countries most public sector organizations rely on inert and inflexible hierarchical systems (Ribot, Agrawal, and Larson 2006), budget inefficiencies (Gupta and Verhoeven 2001), constrained capacities in public information management (Mutula and Wamukoya 2009), and lower levels of public trust in the delivery of services by public institutions (Haque 2001), among others. The implication of these public sector circumstances for SDIs in developing countries is twofold:

- The SDIs rely on spatial data sets that are dispersed over many public sector agencies. There is no national coverage of any spatial data layer, and there are many redundancies in the maintenance processes and the content of the spatial data of public agencies. Therefore, SDIs in developing countries develop in this fragmented context.
- SDI coordination cannot be sufficiently flexible to match the required technological changes with the capacity and willingness of the public

> sector to adapt (Cordeiro and Al-Hawamdeh 2001; Crompvoets and Bregt 2003; Lance and Bassolé 2006). As a result, SDI development in developing countries requires SDI coordination different from that in developed countries.

Uganda's background makes it an interesting and rich case for exploring this mismatch between SDI coordination objectives and its SDI development context. Uganda was an early advocate of systematically coordinating national environmental information (Gowa 2009), and already in 2000 a number of stakeholders had gathered to discuss SDI-related issues with various national SDI frontrunners (Kalande and Ondulo 2006). However, despite this internal momentum in SDI awareness and external donor support enabling local governments, empirical studies report that SDI development is stalled, blaming certain public sector reform processes specifically, such as decentralization (Lwasa 2006; Nasirumbi 2006). The SDI development outcomes are thus strongly influenced or constrained by processes in the public sector. The core question of our research is therefore: How does the Ugandan public sector context explain the differences between SDI coordination objectives and SDI development outcomes?

The structure of the chapter is as follows: First, we present the theoretical framework based on concepts of resource dependency theory (RDT) to explore the Ugandan public sector practices. The next section describes the method of empirical data collection based on this theoretical framework. The subsequent sections analyze and interpret the results in terms of the RDT concepts. We conclude with suggestions on how to improve SDI coordination.

5.2 Theoretical Framework

To understand Uganda's public sector context and how the context may determine or constrain SDI coordination possibilities and alternatives, we use an RDT lens. RDT has proven useful to describe and explain information management processes across organizations. Homburg (1999) refers to the RDT concepts of *power* and *uncertainty* as key components to explain the organizational and political motivation in interorganizational public sector cooperation. Such motivation and willingness factors are crucial in investigating the behavior of public sector actors when confronted with SDI coordination. RDT reasons that organizations engage in relationships with other organizations in order to increase their chances of survival (Montealegre 1998; Pfeffer and Salancik 2003; Patrakosol and Olson 2007). The relationships are built along resource exchanges, which create mutual resource dependencies. The dependencies are a source of *power* for those who own or control the resources and a source of *uncertainty* for those who depend upon the resource (Pfeffer and Salancik 2003).

TABLE 5.1

Relation of Power and Uncertainty to SDI Development Outcomes

Type of Interorganizational Relation	Power	Uncertainty	SDI Development Outcome
Independent of each other's resource, yet jointly dependent on external resource	Power of each partner applied to effectively streamline interorganizational spatial data processes	No uncertainty in obtaining resource → easy cooperation	Efficient, nonredundant SDI
Strongly interdependent on each other's resources	Power of each partner applied for survival of individual organization	High uncertainty in obtaining resource → addressed by risk avoidance	Nonefficient, redundant SDI

Power and uncertainty are thus crucial elements of how public sector organizations work and therefore crucial in explaining how and why organizations accept, adopt, modify, neglect, or reject certain coordination strategies. According to RDT, redundancies of spatial data may emerge if organizations regard the SDI coordination strategy as a threat in their endeavor for survival—for example, because it would raise the dependency on the spatial data resources of one data provider or data consumer only. In this case, organizations would want to avoid the uncertainty arising from being dependent by autonomously creating alternatives for accessing the spatial data.

Furthermore, RDT assumes that power relations (such as public administrative hierarchies, which are often based on financial resource allocations) may determine which spatial data to use and which to collect at which level of authority. On the other hand, if partners do not rely on strong resource dependencies yet have mutual interests in engaging with their environment, they are less uncertain about resource allocations from each other and more likely to team up to serve a mutual interest. The power in this case is used to streamline their interorganizational activities, and less redundancy emerges. Table 5.1 summarizes the relation of power and uncertainty to SDI development outcomes.

5.3 Methods of Data Collection

In the empirical analysis, we focused on how Ugandan public sector organizations handle power differences and uncertainty in the context of spatial data exchange activities, and on where and why redundancy or efficiency results. The data collection started by making an inventory of spatial data producers and custodians in Kampala (the capital) and Entebbe (another major administrative center). These included the member organizations of the National Geographic Information Systems (GIS) task force that was established in 2000, such as the Lands and Surveys Department (LSD), Ministry of Local Government (MOLG), Karamoja Data Center (KDC), and the Uganda Bureau of Statistics (UBOS). Other organizations actively producing and using spatial data include the National Environmental Management Authority (NEMA) and the Makerere University (MUK).

We collected empirical data on "power," "uncertainty," "redundancy," and "efficiency" by a four-step incremental data collection process. This approach provided a gradual insight into the perceptions and behavior of actors. The four steps included:

1. A workshop in Kampala in 2007. This identified 60 key public and private sector SDI stakeholders in Uganda and generated a list of most commonly shared spatial data sets.
2. A survey. From these 60 organizations, 51 representatives responded to a structured questionnaire, which focused on the extent and possible awareness of redundancy and possible efficiency measures (such as standards).
3. Interviews. In both 2007 and 2008, a total of 32 in-depth interviews were held with managerial, technical, and operational staff from 19 public sector organizations. The interviews sought to gauge the perceptions of the influence of power differences and uncertainties on their daily behavior in spatial data sharing in the public sector, as well as to identify possible drivers for redundancy and (in)efficiency in spatial data exchange in the public sector.
4. Focus group discussions. These focus groups discussions, held on October 18, 2007, and October 23, 2008, provided more insight into possible reasons for redundancy, and underlying perceptions of power (differences) and uncertainties. The details of the data collection process, results, and the lists of participants in 2007 are in Nyemera (2008) and in 2008 in Chaminama (2009).

During the interviews and focus group discussions we were attentive to

- The degree to which actors perceive that strategic behavior of their own organization and of other organizations affects their own work with spatial sharing (to gauge how or if *power* affects spatial data processes)
- The degree to which actors perceive how changes in the public sector environment and in internal strategies influence their own views and behavior (to gauge which type of *uncertainty* is present and how this uncertainty affects spatial data processes)

The analysis of these results relied on the axial coding technique of Strauss and Corbin (1998). Axial coding constructs the concepts of power and uncertainty through a coding process of text documents (interview transcripts, open responses in questionnaires, etc.) in a structured (instead of an open) way. One seeks how the basic codes on the concepts power and uncertainty relate to

their causal conditions
the contexts and intervening conditions
the applied action strategies
the consequences

These four contextual elements help to determine how actors perceive concepts and to explain why and when actors refer to the concepts (Strauss and Corbin 1998). This analysis provides the basis for the subsequent interpretation of how and why redundancy occurs as a result of the public sector practice of power difference and uncertainty handling.

5.4 Results

The following results reflect the responses and comments on the extent to which actors perceive the presence of redundancy and efficiency policies and the extent to which they perceive the presence of power and uncertainty.

5.4.1 Presence of Spatial Data Redundancy or Efficiency Policies

The consecutive workshops and interviews identified that many organizations maintain and use similar spatial data sets. Most organizations in Uganda compile spatial data from existing topographic maps, aerial photographs, remotely sensed images, historical records, legal documents, and direct field observations and surveys. Figure 5.1 shows the key data sets that are redundant. "Redundant" in the eyes of respondents meant that more than one organization actively collects and maintains updates of the same database.

More specifically, the survey identified that the "administrative boundaries" (ABs) were the most redundant. The database was regularly copied among stakeholders (which in itself does not reflect redundancy), and, among the 51 organizations surveyed, at least 23 organizations actively maintained and updated modifications of the ABs themselves. Thus, not only the data but also the activities of data maintenance are redundant.

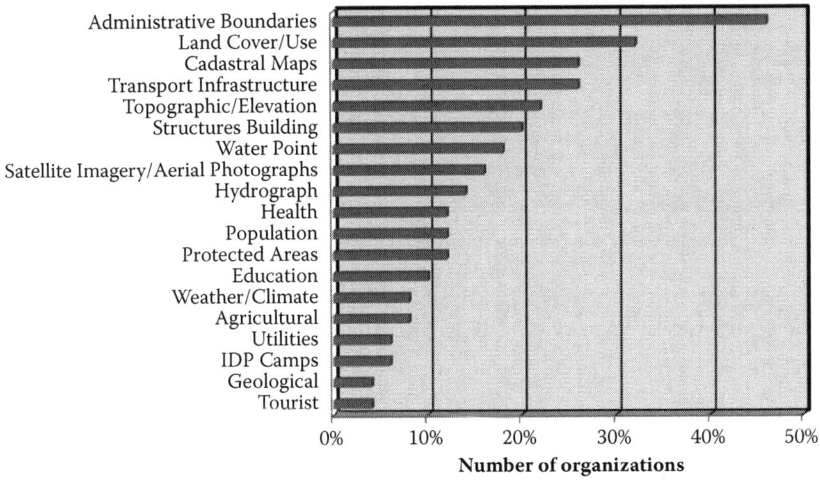

FIGURE 5.1
Most commonly produced and maintained spatial data sets.

Moreover, the inventory discovered that five different organizations produced the AB as unofficial custodians simultaneously: Uganda Bureau of Statistics (UBOS), Ministry of Local Government (MLG), Lands and Survey Department (LSD), Electoral Commission (EC), and the Food and Agriculture Organization (FAO).

5.4.2 Power Presence in Spatial Data Exchange Activities

The respondents argued that power (differences) played a role in three types of cases:

- Directives from national institutions. The National Environmental Act of 1995 establishes, for example, that the National Environmental Authority (NEMA) has the power to "coordinate, monitor and supervise all activities in the field of the environment." NEMA has the power to collect environmental information and all the data needed to collect this information. NEMA could enforce access based on its mandate, but in practice, as one of the respondents in Chaminama (2009) notes, "But I can assure you if NEMA woke up and sa[id], 'I am the boss of water, forestry, and geology; come here and give me information' ... no one w[ould] provide that information." In practice, exercising power does not depend on national directives only.
- Internal organizational mandates and/or vertical public administrative structures. A response such as "At the national level, we don't have a mandate to supervise or to initiate policy for sharing between

ministries. It is strictly at a local government level that we have a mandate" clearly associates power to "authority" and "influence," which are rooted in "mandates." When expanding on the mandates, the same respondent stated, "[O]ur mandate is as much as to really make sure that the information systems follow the right laws and procedures before they can be deployed into local governments"—implying that actors execute mandates within the narrow boundaries of their organizational structures. Only once they are executed may other organizations continue with their own mandates. The respondent perceives that there is no overarching mandate beyond single organizational boundaries.

- Internal assets (as a source of power). The key producers of data acknowledged that the possession of data is a valuable asset or resource to exercise power. "For us as a ministry, because we are a big provider of information, we willed our power" (Chaminama 2009).

5.4.3 Presence of Uncertainty in Spatial Data Exchange Activities

SDI stakeholders appeared highly uncertain about data quality and data access. All respondents to the survey indicated these two issues as problematic. Because of the uncertainty in data quality and data access, they resort to a strategy of creating data backups.

5.5 Analysis—Based on Axial Coding Categories

5.5.1 Redundancy and Efficiency

5.5.1.1 Causal Conditions

During the interviews with five organizations (UBOS, LSD, MLG, EC, and FAO), representatives formulated various reasons why they maintained ABs in addition to other organizations. First, they referred to difficulties in accessing data from the other organizations. Most respondents referred to bureaucratic procedures as a bottleneck for access. To avoid having to go through these procedures and associated annoyances, most staff preferred to collect and maintain the data themselves. They also indicated that the primary mode of access remains physical visits to the offices.

Furthermore, reciprocity is an important element related to interorganizational processes. Chaminama (2009) reports the statement: "If you need information from an organization which does not need information from you they would rather be reluctant to give you." This means that, in reality, interorganizational access and hence redundancy and efficiency are highly personalized. This implies that even if there were a universal access policy, the practice of access is not rooted in adhering to cross-organizational policies, but rather in following interpersonal reciprocities.

Finally, decentralization was perceived as a cause. Through decentralization, the number of districts grew from 36 in 1995 to 76 by mid-2007 (Adams 2006), causing frequent redrawing of administrative boundaries and associated maps. Instead of the concerted spatial data coordination initiatives, the decentralization caused an enormous spur of new and redundant spatial databases at all levels. As a result, the actual Ugandan SDI developed toward redundancy, despite national SDI coordination initiatives aiming to reduce redundancy.

5.5.1.2 Context

Respondents referred to the adoption of GIS in organizations as coinciding with an increase in redundancy. The presence of a large number of donor projects could possibly explain this perception (Lance 2005). When evaluating the five primary custodians of ABs, which include UBOS, MLG, LSD, EC, and FAO, it is apparent that each organization collaborates in separate donor-funded projects. These projects have increased the extent of spatial data redundancy as well as the dependency on donor data specifications (Lance 2009). This also explains the perceived causes of redundancy within the organizations: lack of coordination and lack of uniform standards.

Furthermore, the frequent re-creation of data on administrative boundaries reflects an underlying operational and an institutional dilemma. The donor-funded projects stipulate the data standards, including those for ABs. A pragmatic solution in such cases is to use any AB that happens to be available and accessible, and take the possible deviation from the "official" AB for granted. The fact that most respondents considered the ABs from other organizations outdated makes them rely on the data that they know—hence, the data to which they historically have had access. Furthermore, the institutional dilemma arises directly from the organizational mandates. As most individuals experience serious access problems, they are not aware of how mandates with regard to data provision and access start and end. This results in perceived institutional overlaps.

5.5.1.3 Action Strategies

Representatives from five organizations indicated that spatial data redundancy occurred after they had adopted GIS, which allowed them to exchange spatial data on a countrywide scale. Prior to the rapid growth of GIS technology in Uganda, the technology (mapping equipment and software) was considered too expensive for other organizations to acquire. Currently, the acquisition of GIS has been easier due to cheaper software, Internet-based mapping, and several donor projects.

5.5.1.4 Consequences

The frequent re-creation of new ABs leads to inaccurate, incomplete, and outdated data because recipient customers obtain ABs from different organizations. This introduces a new cycle of redundancy and ultimately reinforces redundancy. Yet one of the hidden effects of redundancy is also the introduction of a functional competition between agencies providing the same data sets. One data user indicated that "with data being available at different locations, it improves the availability, quality, accuracy. It generates special purpose, cheap and affordable data sets." Krause and Douglas (2003) explain this effect of functional competition by a perceived threat that other organizations can do better. As a result, the organization feels it needs to innovate. So, while redundancy is re-created, the same redundancy also generates a cycle of competition that may ultimately generate quality improvement.

5.5.2 Power and Uncertainty

5.5.2.1 Causal Conditions

In Uganda, local government is supposed to be responsible for the implementation of many government services, but this reform has not been followed up with full fiscal decentralization (Williamson 2003). Leaders and managers in ministries are often autocratic (Williamson 2003), leaving technical staff with limited ability to make decisions regarding data infrastructure. In line with this, as mentioned earlier, the majority of the organizations reported not having organization-wide spatial information sharing policies in place (22 out of 27 respondents in 2008). One organization confirmed having a functional spatial information sharing policy in place enforced by the top management, but most respondents reported that informal information dissemination practices were more common than formal information dissemination practices. At the personal level, exchange was considered easier; however, as several respondents indicated, this type of exchange was more prone to personal treatments and personal discretions. This also explains the unwillingness that people perceive when wanting to engage in data sharing activities.

With regard to the causal conditions for uncertainty, many respondents referred to redundancy in relation to heterogeneity in spatial information standards and problems of data quality, compatibility, accuracy, and reliability. Such quality uncertainty has caused organizations to produce their own data sets with quality standards that are applicable to and acceptable for their own use. This option of reproducing already existing data sets is deemed by organizations as an easier task than tailoring and editing data standards to suit their needs.

5.5.2.2 Context

During the interviews, respondents frequently mentioned the lack of uniform standards and a uniform policy applicable for all organizations. One can see this as contextual for the power concept because the primary concern is more likely the lack of enforceability. Even though different laws and regulations are available and there seems a general agreement among the five interviewed organizations that standards and uniform policies would be useful, they all doubted that it would be enforceable. Chaminama (2009) reported the following responses on the influence of government in sharing matters:

> "Influence? No, I think we are sharing, basically our sharing is interinstitutional. I don't think there is any influence."
>
> "I think we are free, unless there is a policy to the contrary, but I think we don't have such policy yet. Government can influence but not now."

The preceding sentence is crucial because it indicates the perception that the problem lies in actual enforcement and compliance, rather than the presence of a policy.

5.5.2.3 Action Strategies

Practically all respondents referred to the lack of coordination and cooperation among public agencies as a primary reason for redundancy and inefficiencies. However, there is sufficient evidence that public agencies do cooperate and even coordinate their cooperation. Chaminama (2009) noted that respondents acknowledged the presence of cooperation, but also noted that many were unhappy about the execution of the cooperation. An interesting comment was that "we developed a cooperation framework that formally did not really work, but is something that recognizes that you are partnering." The lack of cooperation is hereby associated with the lack of being equal partners and is not associated with the willingness to partner. The equality in the partnership is considered more crucial for the success of the partnership than the strategic advantage of the partnership. This is an important implication for SDI coordination because recognizing the level playing field is thus considered more valuable to stakeholders than emphasizing the strategic common advantage.

5.5.2.4 Consequences

The fact that power is an issue in data sharing activities was reflected in a few instances where organizations were pursuing leading roles in the establishment of formal SDIs. This role was taken by specific data collection agencies, such as NEMA, National Forestry Authority, Uganda Bureau of Statistics, and the Lands and Survey Department. These lead agencies aimed to coordinate data sharing at the national level through extending their mandate to include national data sharing. Practically, this power difference was visible through how actors initiate and follow up meetings on data sharing and SDI cooperation.

Often a single organization conducted a meeting and, during such meetings, would convey a general plan hinting at a formal governance structure in which the organizing agency could take a leading role in the SDI process. In none of the cases was a bottom-up or gradual process of coordination developed. This organization would also express the need to create a new (and possibly more neutral) coordinating body and a desire to preserve its autonomy and thus retain control over its data operations and activities. Chaminama (2009) quotes a UBOS representative: "We are trying to lead because we are leading statistical production; we feel we should have an upper hand in GIS in support of our statistical production."

The creation of backup data generates redundancy, yet it also generates the potential and the ability to compare and check data from other organizations. The responses of the five organizations indicated that spatial data redundancy started to occur once GIS was adopted in various organizations. Respondents indicated that they saw an increase in data quality and availability—not despite but because of—data duplication. This seems contradictory. Yet one respondent stated that this redundancy led to cheaper and more affordable data sets than before because they could now rely on subsets of data from different agencies.

As a result, there were more frequent updates of the data throughout the year, given the increased number of providers. Most users appreciated the presence of multiple producers of the same data set because this allowed them to update their data at different times of the year. In addition, two respondents stated that redundancy had created a kind of competition and forced the providing organizations to address inaccuracies. They stated that previously it was difficult to access data sets because these were produced (monopolized) by one organization only—the Lands and Survey Department. However, now the ABs could be accessed from multiple sources.

5.6 Interpretation—Linking Power and Uncertainty Factors to Redundancy and Efficiency

With the preceding analysis, one can go back to the original question of whether the existing power (differences) and uncertainties in the public sector determine the SDI implementation outcome. In the next two subsections, the relation of power and uncertainty to existing redundancy and efficiency is further deduced.

5.6.1 Power as a Cause for Redundancy or Efficiency

The organizational mandate "causes" for redundancy appear to outweigh the efficiency drivers to reduce redundancy. Overlapping organizational mandates (and restricted access to data therein) cause parties to create data sets multiple times. Using the argument of mandates to restrict data access furthermore points to a problem of power more than a problem of efficiency. In itself, this problem has been reported frequently in bureaucracy studies. Wilson (1989) points out that "[n]o agency head is willing to subordinate his or her organization to a procedure that allows other agencies to define tasks or allocate its resources." In other words, spatial data constitute a source of power in the public arena, and organizations are not willing to be under someone else's power.

In addition, the reasons for maintaining redundancy can be found in the administration mandates. The fact that a number of respondents associated the issue of duplication with the issue of data quality confirms that redundancy may be a means to check or prevent administration failures. Landau's seminal essay of 1969 challenged those in the discipline of public management to rethink whether zero redundancy is indeed an appropriate measure of performance or success. Landau (1969) urged:

> There are good grounds for suggesting that efforts to improve public administration by eliminating duplication and overlap would, if successful, produce just the opposite effect. That so many attempts have failed should perhaps alert us to what sociologists would call the "latent function" of this type of redundancy. This possibility alone is sufficient warrant for transforming a precept into a problem. (p. 349)

Built-in redundancy can make an organization more reliable. The redundancy and duplication of ABs can be important for the government of Uganda. They create several parallel interorganizational data processes providing data and data services to the same user population (even though uncoordinated). These parallel processes create a certain competitiveness and provide backup systems in case one of the organizations providing ABs fails in delivering these fundamental data.

While an efficient, nonredundant approach to data development and management is the conceptual vision of SDI, it is problematic from a public sector reality in which power and power differences play a prominent role. Government, by design, is constructed around boundaries. "Boundaries between programs fuel political debate. Boundaries between administrative agencies shape clarity of purpose. Boundaries within agencies, through hierarchy and authority, promote efficiency. These boundaries are essential for defining administrative responsibility and, ultimately, democratic accountability" (Kettl 2001, p. 8).

These ingrained boundaries, by design, are part and parcel of government, including the government in Uganda. In theory, the boundaries should be beneficial, but the practice in spatial data sharing is that the civil servants continue to concentrate on protecting their autonomy rather than thinking holistically about the overall strategy and goals of the government. This creates a fundamental tension when it comes to SDI coordination: the tension between the need to "join up" to achieve cross-agency aims and the institutional drivers that maintain departmentalism. The current power differences are highly influential in resolving this tension.

5.6.2 Uncertainty as a Cause for Redundancy or Efficiency

The uncertainty of organizations in terms of data access and quality may outweigh the efficiency driver of, for example, standardization. The uncertainties limit the willingness of data users to be dependent on a single data custodian. Rather than being dependent, organizations opt for more certain processes of collecting and maintaining data sets themselves. This is a backup mechanism for data unavailability and data unreliability.

5.7 Conclusions

In this chapter, we argued that SDI development can rely not only on coordination objectives emphasizing technological efficiency, but also heavily on the drivers inherent in the public sector context. The current processes of spatial data sharing and the practices of Ugandan data provision have shown that drivers implicit in the public sector context characterize SDI development in Uganda more than the usually touted SDI efficiency drivers. Through the use of resource dependency theory in understanding the public sector context, we were able to identify how power constraints and uncertainty handling influence SDI development. We therefore conclude that a technical, efficiency-oriented SDI agenda can be stymied by the public sector context in which the SDI implementation occurs.

The implication is that SDI strategies and associated implementation plans need to be reoriented to address power and uncertainty realities in each country; they need to address circumstances regarding the role of organizational mandates and the relationships between central and local government (Smit et al. 2009). They need to understand whether or how polices actually have "weight" in institutional and individual public sector behavior, and to assess the context of aid-agency-supported geospatial activity. We need to understand political administrative realities and not assume that technical information management perspectives are easily going to replace deep-seated, institutionalized administrative practices and values.

Furthermore, we argue that data redundancy, against which SDI strategies typically argue, actually may be serving public management purposes in some cases. Some forms of redundancy can increase organizational effectiveness and data reliability and can thus implicitly increase the legitimacy of public sector organizations. As long as the technological resources and the human capacity in Uganda still are limited, the presence of redundancy of spatial data in the public sector is therefore vital for development for Uganda. It allows the public sector to maintain a minimum store of information necessary for its activities. Redundancy may lead to independent quality checks, more exchange and comparison of data, and more freedom to choose data.

Although this research has focused on a single case (Uganda), the implication of the findings is that a shift in the thinking on SDI policies is necessary. Conventional SDI advocacy emphasizes redundancy reduction. This emphasis on efficiency discounts potential benefits of "inefficiency." Hence, an SDI approach more rooted in local public management realities may be more relevant and realistic for developing countries than one rooted in efficiency and data integration assumptions only.

A last general observation with regard to SDI coordination in Uganda is that most ministries have only a few GIS specialists, who end up wearing other hats as well. Their time gets consumed with daily administrative tasks, and coordination is an added burden. Agencies are focused on their primary mission—their business mandates—and not necessarily on the broader policy issue or minimizing redundancy (increasing efficiency) within government. Coordination is not the paramount concern. Complicating matters further, cross-agency coordination typically is an unfunded mandate, with agencies having no resources specifically reserved for or allocated to coordination efforts. Yet cross-agency coordination is a very time- and resource-consuming activity. A shift in coordination rationality is therefore likely to take considerable time and investment of staff and incentives.

References

Adams, M. 2006. Mapping as a tool for planning and coordination in humanitarian operations. *Humanitarian Exchange Magazine*. Issue 36. http://www.odihpn.org/report.asp?id=2851 (last date of access: 4 November 2010).

Astle, H., G. Mulholland, and R. Nyarady. 2006. Bridging the gap towards a standardized cadastral model. *Computers, Environment and Urban Systems* 30:585–599.

Baker, K. S., and C. L. Chandler. 2008. Enabling long-term oceanographic research: Changing data practices, information management strategies and informatics. *Deep-Sea Research Part II—Topical Studies in Oceanography* 55:2132–2142.

Bekkers, V. 2009. Flexible information infrastructures in Dutch e-government collaboration arrangements: Experiences and policy implications. *Government Information Quarterly* 26:60–68.

Buogo, A., and J. J. Chevallier. 1995. Spatial information systems and information integration. *Computers, Environment and Urban Systems* 19:161–170.

Chaminama, N. 2009. Analysis of public sector cooperation and geoinformation sharing: A resource dependence perspective. Master's thesis, ITC, Enschede, p. 83.

Cordeiro, C. M., and S. Al-Hawamdeh. 2001. National information infrastructure and the realization of Singapore IT2000 initiative. *Information Research: An Electronic Journal* 6.

Crompvoets, J., and A. Bregt. 2003. World status of national spatial data clearinghouses. *URISA Journal* 15:43–50.

Fonseca, F. T., M. J. Egenhofer, C. A. Davis, and K. A. V. Borges. 2000. Ontologies and knowledge sharing in urban GIS. *Computers, Environment and Urban Systems* 24:251–272.

Gowa, E. K. 2009. Best practices in environmental information management in Africa. The Uganda case study. Arendal: UNEP/GRID-Arendal.

Gupta, S., and M. Verhoeven. 2001. The efficiency of government expenditure: Experiences from Africa. *Journal of Policy Modeling* 23:433–467.

Haque, M. S. 2001. The diminishing publicness of public service under the current mode of governance. *Public Administration Review* 61:65–82.

Homburg, V. M. F. 1999. The political economy of information management. A theoretical and empirical analysis of decision making regarding interorganizational information systems, University of Groningen.

Kalande, W., and J. D. Ondulo. 2006. Geoinformation policy in East Africa. Paper presented at XXIII FIG Congress, October 8–13, 2006, Munich, Germany.

Kettl, D. F. 2001. Central governments in 2010. A global perspective. Paper prepared for Strategic Thinkers Seminar, Performance and Innovation Unit, Cabinet Office, London.

Krause, G. A., and J. W. Douglas. 2003. Are two heads always better than one? Redundancy, competition and task performance quality in public bureaus. Paper presented at 15th annual conference of the Association of Budgeting and Financial Management, September 18–20, 2003, Washington, DC.

Lance, K., and A. Bassolé. 2006. SDI and national information and communication infrastructure (NICI) integration in Africa. *Information Technology for Development* 12:333–338.

Lance, K. T. 2005. Tracking geospatial investments in Africa. In *Proceedings AfricaGIS 2005*, October 31–November 4, 2005, Pretoria, South Africa, 664–667, EIS-Africa.

———. 2009. How aid agencies transfer geospatial policy and practices to Africa. Paper presented at AfricaGIS 2009 international conference, October 26–30, 2009, Kampala, Uganda.

Landau, M. 1969. Redundancy, rationality, and the problem of duplication and overlap. *Public Administration Review* 29:346–358.

Lwasa, S. 2006. Brief of the national SDI stakeholders' workshop in Uganda, Kampala, Uganda.

Miranda, R., and A. Lerner. 1995. Bureaucracy, organizational redundancy, and the privatization of public services. *Public Administration Review* 55:193–200.

Montealegre, R. 1998. Managing information technology in modernizing "against the odds": Lessons from an organization in a less developed country. *Information and Management* 34:103–116.

Mutula, S., and J. M. Wamukoya. 2009. Public sector information management in east and southern Africa: Implications for FOI, democracy and integrity in government. *International Journal of Information Management* 29:333–341.

Nasirumbi, S. 2006. Towards strategy of spatial data infrastructure development with focus on the private sector involvement: A case study in Uganda. Master's thesis, ITC, Enschede, p. 98.

Nyemera, B. W. 2008. Evaluation of redundancy in the geo-information community in Uganda. Master's thesis, ITC, Enschede, p. 123.

Patrakosol, B., and D. L. Olson. 2007. How interfirm collaboration benefits IT innovation. *Information and Management* 44:53–62.

Pfeffer, J., and G. R. Salancik. 2003. *The external control of organizations: A resource dependence perspective.* Stanford, CA: Stanford University Press.

Rasmussen, J. 1993. Reuse of basic registers—Citizen-service systems operating on basic registers. *Computers, Environment and Urban Systems* 17:297–309.

Ribot, J. C., A. Agrawal, and A. M. Larson. 2006. Recentralizing while decentralizing: How national governments reappropriate forest resources. *World Development* 34:1864–1886.

Smit, J., P. Makanga, K. T. Lance, and W. T. de Vries. 2009. Exploring relationships between municipal and provincial government SDI implementers in South Africa. In *Proceedings of the GSDI 11 World Conference,* June 15–19, 2009, Rotterdam, The Netherlands, 18 pp.

Strauss, A., and J. Corbin. 1998. *Basics of qualitative research. Techniques and procedures for developing grounded theory,* 2nd ed. London: SAGE Publications.

Ting, M. M. 2003. A strategic theory of bureaucratic redundancy. *American Journal of Political Science* 47:274–292.

Vigoda, E. 2002. From responsiveness to collaboration: governance, citizens, and the next generation of public administration. *Public Administration Review* 62:527–540.

Williamson, T. 2003. Targets and results in public sector management: Uganda case study. London: Overseas Development Institute.

Wilson, J. Q. 1989. *Bureaucracy: What government agencies do and why they do it.* New York: Basic Books.

6

Social Network Analysis of the SDI in Flanders

Glenn Vancauwenberghe, Joep Crompvoets, Geert Bouckaert, and Danny Vandenbroucke

CONTENTS

6.1 Introduction: A Network Perspective on SDI

Spatial data infrastructures (SDIs) are developed at different levels of society to facilitate and coordinate the exchange of spatial data (Crompvoets et al. 2004). Many attempts have been made to define and analyze these SDIs, both by practitioners and scientists (Onsrud 1998; Craglia et al. 2003; Masser 2005, 2007; Vandenbroucke and Janssen 2008). Very often SDIs are considered in a hierarchical context in which high levels of SDI (global, multinational, national) build upon lower levels (regional, municipal) (Rajabifard, Feeney, and Williamson 2003). In addition to existing perspectives for defining and analyzing SDIs, a network perspective on SDI was introduced to define and analyze SDI in an operational manner (Vandenbroucke et al. 2009).

The concept of the network captures the dynamic and heterogeneous interactions between a large number of partners. In the network perspective, an SDI is operationalized in terms of organizations that are producing and using spatial data in a shared environment and the flows of spatial data between these organizations. Together, organizations and flows form a network of spatial data exchanges. The organizations are the so-called nodes or actors in this network, while the flows of spatial data between these actors give shape to their relations. An SDI can then be defined as the collection of technological and nontechnological arrangements that give shape to this network of spatial data exchanges between data producers and data users.

The strength of the network perspective in describing and analyzing SDI is twofold. First, the network perspective focuses on the impact that an SDI can have on practices of data access and data exchange. The network perspective enables the analysis of how flows of spatial data between data users and data producers are arranged and facilitated by an SDI. Second, the network perspective allows for an analysis of mutual relationships between different actors and their arrangements within an SDI framework.

Since SDI is defined as a collection of SDI arrangements, producers and users of spatial data are confronted with multiple arrangements, and their access to and use and exchange of spatial data are rarely determined by one single arrangement. Interaction between SDI arrangements at different levels and between levels could be complex, and thus new models are needed to analyze this interaction (Budhathoki and Nedović-Budić 2007). A network perspective on SDI enables us to gain in-depth insights into interactions between different actors and arrangements.

In order to analyze SDI from a network perspective, an appropriate research method is needed. In this study we illustrate how social network analysis can be used as a method to analyze SDI from a network perspective. Social network analysis was pioneered and developed by social scientists as an empirical quantitative method for analyses of social networks (Scott 2000; Knoke and Yang 2008; Wasserman and Faust 2008). We apply this method to analyze SDI in Flanders, Belgium. In the following sections, we introduce social network analysis as a research method. Then we present the case study of SDI in Flanders, including the general background, the process of data collection, and discussion of results. To conclude, we discuss the added value of this network analysis to the analysis of the Flemish SDI in particular and the analysis of SDI in general.

6.2 The Method: Social Network Analysis

In general terms, the concept of a network refers to patterned relationships among individuals, groups, or organizations (Dubini and Aldrich 1991). Any network thus encompasses two indispensable elements: actors and relations (Knoke and Yang 2008). These actors can be all types of social entities—for example, individuals, groups, organizations, or nation-states (Wasserman and Faust 2008). Actors are linked to one another by relational ties. These relational ties can also be of diverse nature. Affective relations, exchange of goods or information, and personal contacts are examples of common types of relations (Knoke and Kuklinski 1982). Social network analysis can be defined as an empirical approach to describe and analyze social structure on the basis of the multiple sets of relations between multiple actors (Keast and Brown 2005).

Social network analysis starts from the statement that actors are part of social systems connecting them to other actors. In determining the behavior of actors, their relations with other actors are more significant than individual characteristics. These relations are the main research object in social network analyses. Two properties of these relations are relevant for further analysis: whether they are directed or nondirected and whether they are dichotomous or valued (Wasserman and Faust 2008). In a directed relation, each tie is directed from one actor to another; in nondirected relations, two actors are connected to each other without a certain direction. In addition, a relation can be dichotomous or valued. In dichotomous relations, information is only collected on the presence or absence of a tie between two actors. In valued relations, ties are further specified with a value as an indication for the strength or frequency of the relation.

Two ways of analyzing can be distinguished within social network analysis: graphical and mathematical (Keast and Brown 2005). Graphical analyses concern the visual representation of networks. These visualizations are based on graphs that represent actors as points or dots; links or relations are expressed as lines connecting these points. Graphs present networks and their relations in a clear and concise way and have strong exploratory and explanatory power (Brandes et al. 1999; Brandes, Kenis, and Raab 2006). Mathematical analyses involve advanced calculations and statistical analyses of the data. The calculations are based on several types of measures, such as the density and centralization of a network, or the centrality of an actor (Haythornthwaite 1996). Both graphical and mathematical analyses can be performed at multiple levels, using specific techniques and measures.

In this chapter we analyze the network of spatial data exchanges in Flanders in a graphical and a mathematical way. First, a graphical representation of the network of spatial data exchanges is given. Mathematical analyses are used to summarize basic characteristics of this network in a number of measures. Three network characteristics are analyzed in order to gain insight into the structure of the Flemish spatial data exchange network: density, centrality, and centralization.

The density of a network refers to the extensiveness of ties between actors within the network (Dubini and Aldrich 1991). Network density is measured by comparing the total number of ties present to the potential number of ties, and it gives us insight into the number of connections within a network. Centrality focuses on the positions of actors within the network and expresses which actors are at the center of the network (Scott 2000). The most central actors are often considered to be the most powerful actors in a network. In general, centrality measures are based on the number of direct and indirect ties actors have to other actors in the network. At the level of the network, one can then determine the global network centralization to express the degree to which a network is centralized around one or more central actors (Freeman 1978).

The objective of this chapter is to evaluate the significance of social network analysis as a methodology in SDI research. Following the definition of Keast and Brown (2005), social network analysis can be seen as an empirical approach to describe and analyze *SDI* on the basis of the multiple sets of *data exchange relations* between multiple *users and producers of spatial data*. The application of social network analysis in SDI research could enable the analysis of how flows of spatial data between data users and data producers are arranged and facilitated by a spatial data infrastructure. The analysis of this network of spatial data exchanges might help us to gain insight into the performance of an SDI and its impact on practices of data access and data exchange. In order to evaluate it, a case study of SDI in Flanders is presented, making use of social network analysis as the research method.

6.3 Background of Flemish SDI

Flanders is the name for the Dutch-speaking northern region of the federal state of Belgium. In 1995 Flanders started to set up a framework for cooperation to develop and implement a sound communication and management system for geographical information: the partnership "GIS-Flanders." Many stakeholders have taken part in the development of GIS-Flanders, including all the departments of the Flemish government, the Flemish public agencies, the provincial authorities, and the municipalities. In 2009 several other public entities such as local police departments and educational institutions were included, and the partnership was reshaped into the partnership "SDI-Flanders" (Flemish Parliament 2009). The aim of the SDI-Flanders is to optimize the production, management, exchange, and use of spatial data in Flanders.

One of the key objectives of SDI-Flanders is to distribute spatial data to all its partners. More than 50 full coverage data sets are available, including street network, Flemish hydrographical atlas, orthophotos, cadastral parcels, digital elevation models, Flemish ecological network areas, land cover, addresses, and soil maps (Crompvoets et al. forthcoming). Within the partnership SDI-Flanders, the regional Agency for Geographical Information Flanders (AGIV) is responsible for the operational coordination and exploitation of the Flemish SDI and operates as its central data distribution hub. The majority of data distributed by the AGIV are produced or collected by other public and private organizations. Some data sets are produced by the AGIV itself. For instance, the AGIV started with the production of a large-scale geographic reference database and a central reference address database (AGIV 2009).

SDI-Flanders can be considered as the central SDI arrangement in Flanders because all regional, provincial, and municipal organizations are members of this partnership. However, within and outside this partnership, several other arrangements are made in order to promote and facilitate the exchange of spatial data. For instance, the distribution of cadastral data to the municipalities falls outside the scope of SDI-Flanders. The Federal Public Service Finance remains in charge of the distribution of its cadastral data to the municipalities.

Other examples of arrangements to facilitate the exchange of spatial data can be found at the level of the provinces. For instance, the provinces of West Flanders and East Flanders both started a partnership with the municipalities in their provinces in 1997; it was aimed at the exchange of spatial data between the province and municipalities. Finally, it is expected that there will be a significant number of more informal or ad hoc data exchanges, in addition to these more formal and institutionalized patterns of data exchange.

The exchange of spatial data in Flanders happens in the context of various formal and informal arrangements between the producers and users of spatial data. As a result, the structure of the Flemish spatial data exchange network is determined by how these arrangements complement each other or mutually interfere. In this chapter, we use social network analysis to examine the structure of the network of spatial data exchanges and the role of different actors and arrangements within this network.

6.4 Social Network Data Collection

Social network data can be collected in various ways. Although the use of questionnaires is the most common method, interviews, observations, and secondary resources can be employed as well (van Duijn and Vermunt 2006). For our empirical study of the exchange of spatial data in Flanders, information was collected through an online questionnaire sent to 508 public organizations. All regional, provincial, and municipal authorities in Flanders were invited to participate in this survey. In addition, a selection of federal organizations, intermunicipal organizations, and harbors was also invited. Private organizations and nonprofit organizations were left out because the focus was on the exchange of spatial data within the public sector (Crompvoets et al. 2009).

The main objective of the study was the identification of spatial data flows among public organizations. In order to identify these data flows in a clear and unambiguous manner, four specific types of spatial data were selected: parcel data, address data, road data, and hydrographic data. For each spatial data type, organizations were asked to indicate whether they used, produced, and/or received these data. In case they received a certain type of data from one or more other organizations, they were requested to identify the suppliers.

Therefore, we made use of a so-called "fixed-choice roster" (Wasserman and Faust 2008). This means that all suppliers could be selected from a list of actors, but the number of choices was restricted to five. For practical reasons, the municipalities in Flanders were collectively considered as one actor (instead of 308 single actors). As a result, data flows could not be registered for each of the municipalities individually. This choice has certain implications for further analyses and should be kept in mind while interpreting the findings.

The questionnaire was returned by 234 public organizations at different administrative levels (Crompvoets et al. 2009). However, some organizations explicitly indicated that they did not make use of spatial data and were excluded from the network analysis. Moreover, only municipal, provincial, regional, and federal organizations were used for analysis. After deletion of cases with missing values, 189 organizations were used for the social network analysis. These organizations constitute the actors in our network.

6.5 An Analysis of the Flemish SDI Network

Social network analysis was used to study the network of spatial data exchanges between 189 spatial data users and producers in Flanders. The method was applied to measure the density and the centralization of the network and the centrality of the actors in this network. We consider this network as a single relational network of spatial data exchanges, meaning that the analysis does not make a distinction between the four types of data. All data flows were considered as the same type of relationship.

It is, however, important to note that the type of spatial data is a significant factor in determining the structure of a network because practices of spatial data exchange vary for the different types of data. Therefore, additional analyses need to be performed to examine the potential impact of the types of data on the structure of the network. In addition, we should not neglect the fact that many other types of spatial data are distributed and exchanged in Flanders. Our network analysis of the exchange of four specific types thus offers just a partial image of the SDI. Caution should be exercised in generalizing our research findings to the SDI in Flanders as a whole.

Our network of spatial data exchanges was analyzed in a graphical and a mathematical manner. A graphical representation of the SDI network is presented in Figure 6.1. In this figure, all organizations are represented as

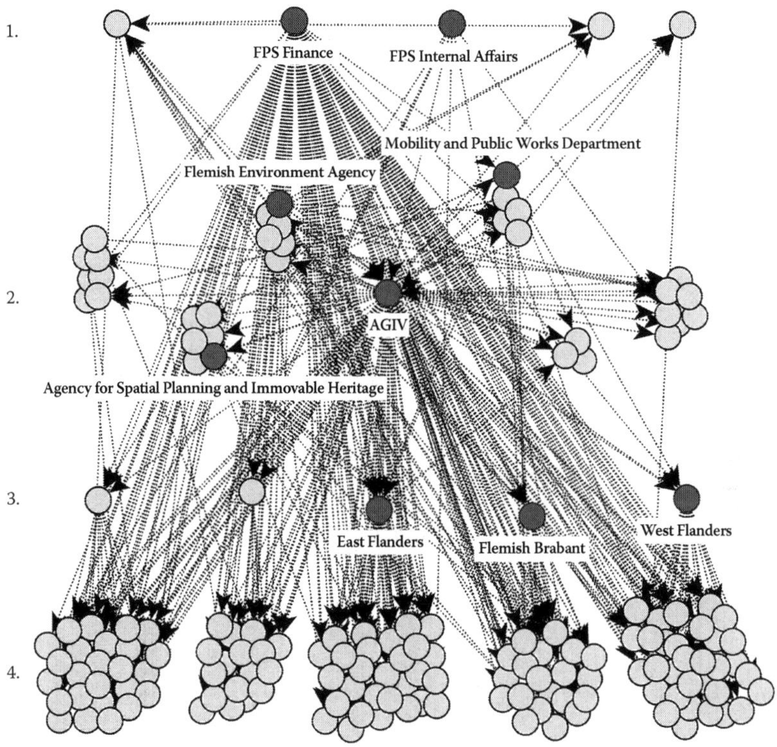

FIGURE 6.1
The Flemish spatial data exchange network.

nodes, while their relational ties are represented by lines.* The actors in this network are positioned by their administrative levels. Four levels can be distinguished: (1) the Belgian federal organizations (at the top), (2) the organizations of the Flemish public administration, (3) the five provinces, and (4) the municipalities (at the bottom row).

Between these organizations, 410 lines can be discerned. These lines refer to the exchange of at least one type of spatial data. Even organizations that exchange more than one type of data are linked with only one line in Figure 6.1. However, if we consider each type of data as a distinct relational tie and allow organizations to be connected by multiple ties, the total number of ties becomes 713. The most central actors in the network are marked and identified. The concept of centrality and its measurement is discussed later in this chapter. First, we focus on the two basic characteristics of the network as a whole: density and centralization.

* At the federal level, Figure 6.1 only presents the five most central actors. At the municipal level, the actors are grouped according to the province in which they are located.

6.5.1 Network Density and Centralization

Several measures of density and centralization of the studied network are presented in Table 6.1. These measures can help us to interpret the visual representation of the network. A first characteristic of all networks is the network's density. Density refers to the number of ties within the network and is expressed by the proportion of all possible ties that are present (Dubini and Aldrich 1991). Table 6.1 distinguishes the density of the single-datum network and the density of the multidata network.

The single-datum network neglects the fact that organizations can exchange multiple types of data. Each relational tie refers to the exchange of at least one of the four types of spatial data. If we consider the Flemish SDI network in this way, a maximum number of 35,532 ties can be reached. Because 410 of the possible number of 35,532 exchanges are realized, the density of the network is 0.012. If we now consider our network as a multidata network with four possible relations, the number of possible ties is multiplied by four. However, each pair of actors can now be connected in eight different ways (four types of data in two directions). In the survey, 713 data relations were registered, leading to a density of 0.005.

The interpretation of both scores is complicated because no comparison can be made with other data exchange networks. In addition, the problem of how to determine the ideal density of an SDI network is still open. However, some basic interpretations of our density scores can be made. If our network had a density of zero, no data would be exchanged between the different actors in the network. Because SDIs are developed to promote the exchange of spatial data, a density close to zero indicates an ineffective—or absent—SDI. On the other hand, a density of one would mean that each organization exchanges data with all other organizations. It can be argued that the SDI network would then be saturated, and the number of redundant data exchanges would be high.

An optimization of the Flemish SDI can change the density of the SDI network in both directions. For example, the analysis revealed that 12 organizations in our analysis had no data ties to other organizations and were thus isolated from the network. These organizations only make use

TABLE 6.1

Density and Centralization of the Flemish SDI Network

	Flemish SDI Network
Network Density	
Single-datum network	0.012
Multidata network	0.005
Network Centralization	
Incoming data flows	1.08%
Outgoing data flows	29.07%

of self-produced data and do not share data with other organizations. The connection of these organizations to the SDI network would raise the density of the network. On the other hand, we see that several organizations are using identical or similar data from several data suppliers. It can be argued that some data exchanges are redundant. The elimination of these exchanges would decrease the density of the network.

The second basic characteristic of a network is the network's centralization. "Centralization" refers to the extent to which a network is structured around one or a few actors (Freeman 1978). In fully centralized networks (centralization degree of 100%), all relational ties are situated around one single actor. In a fully decentralized network, the number of relational ties is equal between all nodes.

In Table 6.1, we report the centralization of the spatial data exchange network in Flanders for incoming and outgoing data flows. For that which concerns the incoming data flows, the network is centralized for 1%. In terms of outgoing data flows, the centralization of the network is 29%. These percentages express the degree of inequality or variance in our network as a percentage of that of a fully centralized network (Hanneman and Riddle 2005). Based on these scores, we could argue that the reuse of spatial data in Flanders is strongly scattered around all the involved actors, while the distribution of data is centralized around a limited number of actors.

If we take a look at the literature on centralization and decentralization within the public sector (Pollitt 2005), arguments can be found in favor of as well as against a centralized distribution of spatial data. To begin, it can be argued that centralization enables (public) organizations to benefit from economies of scale. In addition, it encourages specialization because it enables organizations to lead a critical mass of experts. Finally, centralization also leads to greater equity because all clients receive the same service.

Despite these benefits, arguments also can be made in favor of a decentralized distribution of spatial data. It can be stated that a decentralized approach increases recognition of subcentral authorities as partners in the development of an SDI. In that way, decentralization means that freedom is given to subcentral authorities to develop their own policy. Decentralization also leads to faster and thus more efficient decision making and can be promoted in the context of spreading power and responsibility.

6.5.2 Actor Centrality

While the concept of centralization is used to refer to a particular characteristic of the network as a whole, (actor) centrality is concerned with the position of a certain actor within this network. The centrality of an actor can be considered in several ways. The question of which actor is the most important or most powerful actor in a network can be answered from different perspectives. From a *degree perspective* on centrality, the actor with the largest number of relational ties is considered to be most central.

It can be argued that actors with many incoming ties have many opportunities to satisfy their needs, while actors with many outgoing ties are able to influence the behavior of many others. The centralization degrees mentioned earlier in this chapter were based on this degree perspective on centralization.

However, degree centrality is often criticized because it only takes into account the direct ties an actor has and neglects the fact that actors can be connected by indirect ties. Therefore, a *closeness perspective* on centrality is often used as an alternative for the degree perspective. Closeness centrality takes into account both direct and indirect ties by focusing on the distance of an actor from all other actors in the network. Actors that are closer to many other actors can then be regarded as central—or powerful—actors (Hanneman and Riddle 2005).

These different perspectives on centrality can be expressed in two measures. Table 6.2 presents the five most central actors in the Flemish SDI network, based on a computation of their out-degree. This measure refers to the total number of outgoing ties from each actor. All four types of data are taken into account (i.e., each data exchange is regarded as a unique relational tie). With an out-degree of 328, the AGIV can be seen as the most central actor in the SDI network. The AGIV is responsible for 328 of the total number of 713, or 45%, of registered data flows.

The second most central in the network, from the perspective of the out-degree, is the Federal Public Service (FPS) Finance. The FPS Finance has an out-degree of 149. Other central actors in the network are the provinces—in particular, the province of East Flanders (49) and the province of West Flanders (13). Just like the AGIV and the FPS Finance, they can be considered as important actors in the Flemish spatial data exchange network because they are directly linked to a large number of other organizations and thus have an impact on the spatial data use in these organizations.

If we take a look at the centrality of actors from the perspective of closeness, the result is somewhat different. One measure to express the closeness centrality of actors is "reach centrality" (Hanneman and Riddle 2005). Reach centrality focuses on the number of steps each actor needs to reach all other actors. The maximum score is equal to the number of nodes and is achieved when all actors are only one step away. This score becomes smaller as actors are two, three, or more steps away.* The five most central actors based on the computation of their reach centrality are listed in Table 6.3.

Based on this measure, the FPS Finance and the AGIV are the most central actors in the network. However, three new central actors emerge: the Flemish Environment Agency, the Flemish Mobility and Public Works Department, and the Federal Public Service Internal Affairs. The graphical image of our network (Figure 6.1) can be used to explain this situation. The figure demonstrates that these three organizations supply data to the AGIV and that

* Actors that are two steps away are weighted 1/2, actors that are three steps away 1/3, etc.

TABLE 6.2

Five Most Central Actors in the Network Measured by "Out-Degree Centrality"

	Out-Degree Centrality
1. Agency for Geographical Information Flanders (AGIV)	328
2. Federal Public Service Finance	149
3. Province of East Flanders	49
4. Province of West Flanders	32
5. Flemish Agency for Spatial Planning and Immovable Heritage	13
Flemish Environment Agency	13
Province of Flemish-Brabant	13

TABLE 6.3

Five Most Central Actors in the Network Measured by "Reach Centrality"

	Reach Centrality
1. Federal Public Service Finance	140
2. Agency for Geographical Information Flanders (AGIV)	137
3. Flemish Environment Agency	80
4. Flemish Mobility and Public Works Department	78
5. Federal Public Service Internal Affairs	76

AGIV distributes these data to other organizations. More specifically, the Flemish Environment Agency and the Flemish Mobility and Public Works Department are the key producers of hydrographic data in Flanders, while the address data of FPS Internal Affairs are used by the AGIV in the production of the central reference address database.

These organizations may not have a large number of direct connections to other actors in the Flemish SDI because their data are distributed to other organizations by the AGIV. Despite the absence of direct links, each of these organizations should be regarded as an important actor in the SDI. In that way, the "reach" perspective on network centrality offers additional insight into the SDI. The most central actors in this SDI are not only the organizations that distribute a large number of data, but also the organizations that are at the source of these data sets.

6.5.3 Summary

The density and centralization measures and the analysis of positions of different actors provide insight into how different SDI arrangements give structure to a network of spatial data exchanges in Flanders. The formal analysis of SDI strongly emphasizes the central role of the partnership SDI-Flanders in determining the access to and exchange of spatial data. From a network perspective, SDI-Flanders shapes the role of the AGIV as central data distribution node.

The network analysis reveals that almost half of the spatial data flows in Flanders start from the AGIV. In addition, the AGIV also has a large indirect impact on the SDI because AGIV data are redistributed by several other organizations. The significance of SDI-Flanders to the exchange of spatial data is also expressed by the central position of several data producers. Although there are relatively few direct links between these producers and the users of their data, these data producers indirectly determine the structure of the data exchange network through the distribution of their data by the AGIV.

The provinces in Flanders can also be considered as important actors in the SDI network. They bring together data from different sources and distribute these data in an integrated way to the municipalities. Almost 15% of all analyzed data flows are initiated by the provinces. However, major differences can be found in the behavior and position between these five provinces. While the provinces of East Flanders and West Flanders have a very active data distribution policy, the distribution of spatial data in the other provinces is rather limited.

Another important actor is the FPS Finance—the federal agency responsible for the production of cadastral maps. The FPS Finance gives structure to the exchange of spatial data in both a direct and indirect way. It is in charge of the distribution of these data to the municipalities. The distribution to other public organizations in Flanders falls within the scope of SDI-Flanders and is organized by the AGIV. However, several direct data exchanges between the FPS Finance and organizations at the regional and provincial levels can be distinguished in Figure 6.1.

The direct impact of the FPS Finance is thus not limited to the distribution of its data to the municipalities. These cadastral data exchanges from the FPS Finance to Flemish public administrations or provinces are the result of bilateral agreements between two organizations. Similarly, many other individual or informal data exchanges within and between administrative levels can be observed. If we total the out-degree centrality scores of the AGIV, the FPS Finance, and the five provinces, we see that approximately 80% of the analyzed data flows can directly be linked to one of these actors. This means that one fifth of the data flows in our analysis cannot be explained by the formal arrangements that were considered in this chapter. One fifth of the network of spatial data exchanges is thus determined by other formal and less formal arrangements.

6.6 Conclusion

The key objective of this chapter was to evaluate the significance of social network analysis as a methodology for SDI research. Therefore, the methodology was applied to analyze the SDI in Flanders. We consider the SDI as the collection of arrangements that give shape to a network of spatial data exchanges between users and producers of spatial data. The starting point for our analysis was a formal-legalistic analysis of the Flemish SDI. On the basis of this analysis, three main SDI arrangements are distinguished: the partnership SDI-Flanders, the provincial SDI initiatives, and provision of cadastral data by the Federal Public Service Finance.

This chapter demonstrates that social network analysis can provide insight into the impact of each of these arrangements on the access to and the exchange of spatial data in Flanders. Of the analyzed data, 80% are facilitated by one of these arrangements. In addition, the chapter illustrates how social network analysis can be used to analyze the interactions between these different arrangements. For instance, the graphical network analysis of the SDI shows that, to a certain extent, these arrangements are complementary to each other—in the sense that they redistribute each other's data to specific groups of users. The analysis demonstrates that the Flemish SDI is partly hierarchically organized because data from higher administrative levels are distributed by lower administrative levels. It is worth noting that social network analysis can be used to reveal the hierarchical characteristics of an SDI in an empirical, quantitative manner.

Nevertheless, the SDI in Flanders cannot be considered as completely hierarchical. For instance, the provincial redistribution of spatial data is restricted to a limited number of provinces. In some cases the provincial SDI level is missing. To a large extent, the distribution of spatial data is centrally organized because the AGIV provides spatial data to all the administrative levels.

Social network analysis is useful in demonstrating the overlap between the different SDI arrangements. The method allows us to illustrate that identical or similar types of spatial data are provided by multiple arrangements to the same users. Finally, social network analysis also takes into account the more informal exchanges of spatial data. In our analysis, one fifth of all data flows could not be linked directly to one of the formal arrangements we considered.

In general, the significance of social network analysis for SDI research lies in its focus on different forms of SDI arrangements: formal as well as informal arrangements and hierarchical as well as nonhierarchical arrangements. In addition, social network analysis enables the analysis of how each of these arrangements complements and interferes with the other. However, it is important to mention that there are some potential pitfalls in the application of social network analysis as a research method. To a large extent, these pitfalls can be artifacts of the process of data collection. As SDIs are analyzed from the viewpoint of organizations and their data relations, significant efforts are needed to collect data on both the organizations and their data relations.

An important concern in social network analysis is the issue of nonresponse or noncollected data. Missing data lead to holes in the network and distort it (Keast and Brown 2005). In addition, attention is needed for the issue of accuracy of self-reported data since relational data are often collected by having people report on their own interactions or the interactions of the organization they are representing (Wasserman and Faust 2008). Therefore, consideration must be given to the selection of individuals who have knowledge of the information needed.

Being aware of these potential pitfalls, the application of social network analysis in SDI research can lead to useful and interesting insights into the development and the use of SDIs. Subsequently, these insights can be extended, complemented, or deepened through the use of other approaches or methodologies. In analyzing SDI from a network perspective, special importance is given to how the current infrastructure gives shape to a network of spatial data exchange between different organizations. As such, the network perspective can serve as a valuable starting point for the further development of the infrastructure.

The main objective of this development should be to optimize the network of spatial data exchanges. However, some important questions still remain unanswered. For example, what is the optimal form of a network of spatial data exchanges? And how can different technological and nontechnological arrangements be used in order to optimize this network? In order to find answers to these basic questions, the contribution of other research approaches is still needed.

References

AGIV. 2009. What is the FGIA dealing with? Products—AGIV—Agentschap voor Geografische Informatie Vlaanderen. http://www.agiv.be/gis/organisatie/?catid=147 (accessed October 7, 2009).

Brandes, U., P. Kenis, and J. Raab. 2006. Explanation through network visualization. *Methodology* 2 (1): 16–23.

Brandes, U., P. Kenis, J. Raab, V. Schneider, and D. Wagner. 1999. Explorations into the visualization of policy networks. *Journal of Theoretical Politics* 11 (1): 75–106.

Budhathoki, N., and Z. Nedović-Budić. 2007. Expanding the spatial data infrastructure knowledge base. In *Research and theory in advancing spatial data infrastructure concepts*, ed. H. Onsrud, 7–31. Redlands, CA: ESRI Press.

Craglia, M. et al., eds. 2003. GI in the wider Europe. http://www.ec-gis.org/ginie/doc/ginie_book.pdf (accessed February 14, 2009).

Crompvoets, J., A. Bregt, A. Rajabifard, and I. Williamson. 2004. Assessing the worldwide developments of national spatial data clearinghouses. *International Journal of Geographical Information Science* 18 (7): 665–689.

Crompvoets, J., E. Dessers, T. Geudens, et al. 2009. Het Vlaamse GDI—Netwerk. Een kwantitatieve verkenning van het gebruik en de uitwisseling van geodata in Vlaanderen.

Crompvoets, J., G. Vancauwenberghe, G. Bouckaert, and D. Vandenbroucke. Forthcoming. Practices to develop spatial data infrastructures as support to e-government. In *Practical studies in e-government*, ed. S. Assar, I. Boughzala, and I. Boydens. New York: John Wiley & Sons.

Dubini, P., and H. Aldrich. 1991. Personal and extended networks are central to the entrepreneurial process. *Journal of Business Venturing* 6 (5): 305–313.

Flemish Parliament. 2009. Decree on the geographical data infrastructure Flanders (GDI decree). *Belgian Moniteur*, April 28, 2009.

Freeman, L. C. 1978. Centrality in social networks. *Social Networks* 1:215–241.

Hanneman, R. A., and M. Riddle. 2005. Introduction to social network methods. Riverside, CA: University of California. http://www.faculty.ucr.edu/~hanneman/nettext (accessed August 7, 2008).

Haythornthwaite, C. 1996. Social network analysis: An approach and technique for the study of information exchange. *Library and Information Science Research* 18:323–342.

Keast, R., and K. Brown. 2005. The network approach to evaluation: Uncovering patterns, possibilities and pitfalls. Paper presented at Australasian Evaluation Society International Conference. South Bank, Brisbane.

Knoke, D., and J. H. Kuklinski. 1982. *Network analysis*. Newbury Park, CA: SAGE Publications.

Knoke, D., and S. Yang. 2008. *Social network analysis*. Los Angeles, CA: SAGE Publications.

Masser, I. 2005. *GIS worlds—Creating spatial data infrastructures*. Redlands, CA: ESRI Press.

________. 2007. *Building European spatial data infrastructures*. Redlands, CA: ESRI Press.

Onsrud, H. J. 1998. Compiled responses by questions for selected questions. Survey of national and regional spatial data infrastructure activity around the globe. Global Spatial Data Infrastructure Association. http://www.spatial.maine.edu/~onsrud/GSDI.htm (accessed October 22, 2009).

Pollitt, C. 2005. Decentralization. In *The Oxford handbook of public management*, ed. E. Ferlie, L. E. Lynn, and C. Pollitt, 371–397. Oxford, England: Oxford University Press.

Rajabifard, A., M.-E. F. Feeney, and I. P. Williamson. 2003. Spatial data infrastructures: Concept, nature and SDI hierarchy. In *Developing spatial data infrastructures. From concept to reality*, ed. I. P. Williamson, A. Rajabifard, and M.-E. F. Feeney, 17–40. London: Taylor & Francis.

Scott, J. 2000. *Social network analysis: A handbook*. London: SAGE Publications.

Vandenbroucke, D., J. Crompvoets, G. Vancauwenberghe, E. Dessers, and J. Van Orshoven. 2009. A network perspective on spatial data infrastructures: Application to the sub-national SDI of Flanders (Belgium). *Transactions in GIS* 13 (1): 105–122.

Vandenbroucke, D., and K. Janssen. 2008. *Spatial data infrastructures in Europe: State of play 2007*. Leuven: Belgium.

van Duijn, M. A. J., and J. K. Vermunt. 2006. What is special about social network analysis? *Methodology: European Journal of Research Methods for the Behavioral and Social Sciences* 2 (1): 2–6.

Wasserman, S., and K. Faust. 2008. *Social network analysis: Methods and applications*. Cambridge, England: Cambridge University Press.

7

Thinking in Circles: How National Geo-Information Infrastructures Cannot Escape from the Temptation of Technology

Henk Koerten and Marcel Veenswijk

CONTENTS

7.1 Introduction

National facilities for collection and dissemination of location-based (geo)-information (GI), now created worldwide, are expected to make government and business organizations more effective (Crompvoets 2006). Within organizations, geo-information is often managed with geographical information systems (GIS) technology and between organizations through national geo-information infrastructures (NGIIs) (Rajabifard and Williamson 2001; Nebert 2004; De Bree and Rajabifard 2005; Masser 2005).

The implementers of NGIIs are inclined to value organizational aspects of its development using design rules borrowed from political science, economics, and management science (Koerten 2007). Accordingly, NGII researchers focus on best practices, organization models, and planning (Rajabifard and Williamson 2001; Warnest et al. 2003; Masser 2005; Warnest, Rajabifard, and Williamson 2005; Obermeyer and Pinto 2008; Box and Rajabifard 2009).

However, some researchers maintain that such an approach is limited. In order to get a better understanding of implementation processes, they focus on nontechnological aspects in a nonprescriptive way, for which they believe alternative ontologies and epistemologies are needed (Harvey 2001; Georgiadou, Puri, and Sahay 2005; Crompvoets et al. 2008; Georgiadou and Harvey 2007). Although some research has been conducted in this vein (Martin 2000; Harvey 2001; Georgiadou and Homburg 2008; Lance, Georgiadou, and Bregt 2009), mainstream NGII research remains prescriptive (Budhathoki and Nedović-Budić 2007; Crompvoets et al. 2008).

Budhathoki and Nedović-Budić (2007) argue that research on geo-information infrastructures is not well grounded in theory and that NGIIs are not effective because a coherent user perspective is lacking; this keeps us from finding answers to problems with NGII implementation. We follow the authors, who attempt to find new ontologies and epistemologies applicable to NGII research (Georgiadou and Harvey 2007) and propose a narrative approach that allows us to focus on how practitioners engage in NGII implementation. We examine how the NGII is narrated in meetings, interviews, and policy documents in order to broaden our understanding about its conceptualization and usage. Our research question is whether *narrative analysis* has value for understanding NGII implementation. Secondary questions include: How do technological and organizational aspects interact with each other? How are goals and results perceived over time?

We use the narrative approach in an in-depth ethnographic case study of a Dutch NGII implementation project called "geo-portals." The project was originally intended to realize a part of the Dutch NGII by disclosing governmental geo-information in a thematically organized way. After a theoretical elaboration, we give an account of the research methodology. An analysis of the project in theoretical terms is followed by concluding remarks.

The Space for Geo-information Program (SGI or, in Dutch, *Ruimte voor Geoinformatie*) ran from early 2005 until 2008, with a budget of €40,000,000 to provide grants for innovative projects contributing to the Dutch NGII. Sector-wide discussions on SGI program goals led to the geo-portals project. The project had a €2,000,000 budget and aimed to establish a network of geo-portals for disclosing geo-data. We conducted an ethnography of the project's initiation, course, and outcome.

7.2 The Narrative Analysis Approach to Research

The objective of the research presented in this chapter is to develop and apply a narrative approach for NGII research. We use interpretation, creation of meaning, and sense-making as the lead concepts of an interpretive method (Polkinghorne 1988; Gergen 1994; Hatch and Yanow 2003). Interpretive methods focused on meaning can be described as a discourse-oriented "linguistic turn" and a story-oriented "narrative turn" (Verduijn 2007).

Be it in either a primitive sign system or sophisticated language, people use language to communicate. In order to study organizations, Grant, Keenoy, and Oswick (1998) conceptualized organizational discourse connecting language to time and space as "languages and symbolic media we employ to describe, represent, interpret and theorize what we take to be the facticity of organizational life" (p. 1). Discursive research is more aimed at how developments over time are brought into a meaningful whole, instead of registering experience as such (Burrell 2000; Shenhav and Weitz 2000). Discourse is seen as an "alternative way of describing, analyzing, and theorizing processes and practices constituting the 'organization'" (Grant 2003, p. 5). Put differently, discourse seems to focus on how dynamics of talk and texts create something static (Reed 2000).

The dynamic character of organizational practice has invoked interest for other linguistic aspects than text alone, such as metaphor, stories, novels, rituals, rhetoric, language games, drama, conversations, emotions, and sense-making (Grant et al. 1998). Grounded in literary criticism, the narrative turn aims to delineate stories and storylines rather than texts (Frye 1957; Burke 1969; Gabriel 2000). Narrative is ubiquitous; we live in a "storified" world where people tell each other stories (Gabriel 2000). Stories are used for exchanging information and meaning about experiences. In giving an account of what happened, people make a selection of events, telling them in a favorable manner.

Either for single use or for retelling, stories get altered and are a frame of reference for future stories and actions. They become narratives, loosely connected to the original (Boje 2001; Tesselaar et al. 2008) and universal images, culminating in identity creation (Beech and Huxham 2003). From a manager to a company car, identities are created by storytelling, leading to continuously reconstructed narratives, being prominent or latent, conscious or unconscious, real or imagined (Boje 2001). In the search for a clear overall picture, blanks are filled with fantasies that function as experiences (Ricoeur 1973; Bruner 1991).

Humans only notice changes in a narrative when it is reduced to a series of instances (Bergson 1946; Burrell 1992, 2000). Once multiple stories start to live a life of their own, they grow out to be narratives, sometimes poorly, or maybe even ill connected, to the stories that brought them to life (Boje 2001). Narratives can be universal, constituting images of all kinds of aspects of society. They refer to cultures of all kinds of groups of people, culminating in creating identities made up from social categories (Beech and Huxham 2003). Identities like a carpenter, a manager, or a woman are thus created by storytelling, culminating in narratives.

Likewise, narratives conceptualize nonhuman identities like office, workplace, the Internet, or company car. These are not fixed concepts, but rather narratively constructed images, which are continuously constructed and therefore subject to change. The framework used in this research conceptualizes the creation and maintenance of stable narratives about actors and

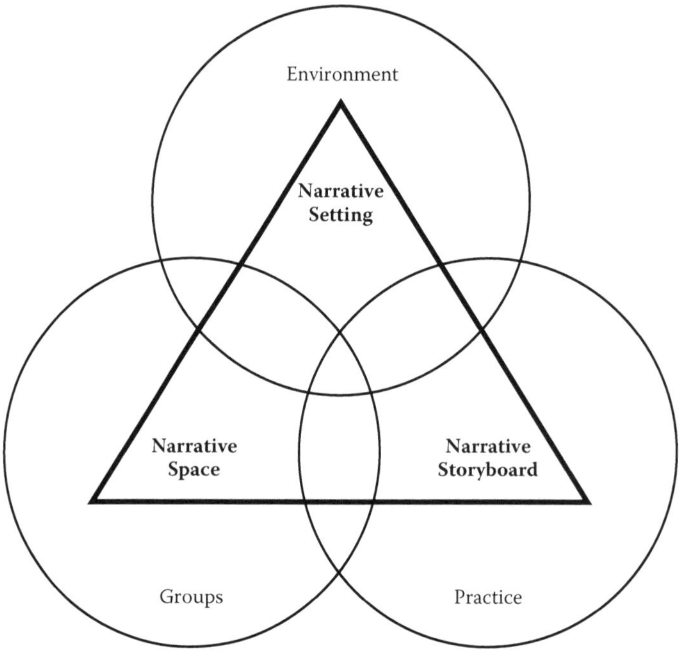

FIGURE 7.1
Theoretical focus.

their actions in terms of narrative setting, narrative space, and narrative storyboard, respectively (Figure 7.1) (Burke 1969; Harré 1976).

The *narrative setting* conceptualizes the environment in terms of time, territory, and technology. A location is enacted, using images from the past, present, and future, from the local community to the global environment (Douglas 1986; Lefebvre 1991; Scott 1995). Entailing the physical environment (Yanow 1995; Gastelaars 2008) and technology (Orlikowski 2000), locations may have different meanings relating to tangible and nontangible aspects (Schneider 1987; Lefebvre 1991; Weick 1995).

An intangible software program, used through a tangible computer, may have a fundamental impact on how things are done (Orlikowski 2000). Technology is shaped through subjective, partial, and distorted images of applications (Bijker 1995; Orlikowski 2007) linked to time and territory (Burrell 1992). This process often results in a relatively stable image of the environment, which will only be redefined when it becomes untenable.

Narrative spaces are departments, organizations, professions, religions, or any other configurations with human actors. As mental "zoning plans" for enacted human groups, narrative spaces invoke action or, conversely, create a deadlock or cease-fire. Because actors are always linked to multiple narrative spaces, they may form quite complex combinations not linked to formal organizational structures (Lipsky 1980; Douglas 1986; Schein 1996).

Change is conceptualized as moving from one form of stability to another (Barley 1990; Bartunek 2004; Ybema 2008).

Narrative storyboards are the bedrock of human action, providing predefined scripts in a world that is made up of a constant flow of events and creating fixed recipes, based on past, present, and future actions (Bergson 1946; Berger and Luckmann 1967; Weick 1995). People feel uncomfortable when mostly unwritten rules are not appropriately applied (Garfinkel 1984). Storyboards provide a narrative to move from one state of affairs to the other, linking action to the narrative setting and narrative spaces. Apprentices may use storyboards to learn the accepted ways of doing (Wenger 1998), moving from chaos to order (Latour and Woolgar 1986), or to be of help to people to know what to expect (Czarniawska-Joerges 1998). Storyboards provide standard ways to enact action, either to be followed through or to be used for sense-making of action (Boje 1995).

The narrative approach was used in this research to analyze the development of the Dutch geo-portals. Empirical evidence was obtained through the ethnographic study pursued by one researcher as participant observer. Ethnographers have to be convincingly authentic ("been there") and plausible (relevant to the reader) and need to engage in critical analysis (Golden-Biddle and Locke 1993). In order to do so, this research project followed writing conventions developed by Watson (2000) and extended by Duijnhoven (2008) concerning the transfer of field notes into convincing and authentic texts.

The researcher monitored the geo-portal project during its existence as a member of the project committee. Monthly management meetings, two brainstorm sessions, and four workshops were attended and also 22 interviews were conducted with key persons, during both start-up and conclusion phases of the project. Relevant documents and video footage were also analyzed. In the next section, we present the excerpts from our interviews and field notes, which are selected and condensed to representations of the typical form of a discussion or commentary concerning a particular topic. The research materials reveal the narratives and illustrate how projects function as arenas for narratives.

7.3 The Geo-Portals Case Description

This section provides a detailed description of three phases of the geo-portals project: outset, middle section, and outcome.

7.3.1 Outset: Getting Geo-Portals Started

The space for geo-information (SGI) program started in 2002 with the intention to boost innovation through a large consortium of governmental and nongovernmental organizations (RAVI 2003). It was argued that to develop effective policies, government would need complex information about society, preferably ordered spatially and accessible through a spatial data infrastructure (NGII). SGI organized "brokering and bargaining days" on which representatives of organizations from the GI sector were invited to generate project ideas. It was in this context that the concept of geo-portals emerged. Here are some typical observations of those in attendance:

> SGI mobilized the field. Some 25 ideas were identified as potentially successful. In the end, these ideas were connected to organizations; it was just one big dating show. The geo-portals concept just came out of a plenary discussion. Then the moderator asked which organizations were willing to participate. Representatives of interested organizations raised their hands, as did I. So, all of a sudden I was an initiating member of an instantly formed club of enthusiastic people who wanted to disclose geo-information through portals. That the overarching concept of geo-portals should be *liberty united* was obvious from the outset. A central, top-down organization was totally out of the question.

In October 2002, representatives from 13 organizations presented an initial proposal, which envisioned thematically categorized, color-coded portals—for example, red for built environment and green for nature and agriculture (Schmidt and Nieuwenhuis 2002). The project plan at the outset was to provide Web-based static images based on proven technology, standards, and architecture (Hoogerwerf and Vermeij 2005). Representatives of the participating organizations, uncertain about the essence of the project, discussed how to proceed:

> A: We need to be clear about standards. It is obvious that we use the most recent and commonly used standards. We are not going to use any standard that has not proved to be useful.
> B: I agree on that. If nobody objects, we should proceed to the next topic, and that is user orientation. How can we be demand driven?
> C: First and foremost we need to disclose our data in a way that they can be readily found, in a user-friendly format. So, we need to use the proper standards.
> B: I agree. We need to use proper standards, those that are widely accepted.
> A: Now we have agreed on how to settle the standards issue, we are discussing standards again.

The motto was "liberty united": a network of portals, established by autonomous organizations, using a minimal set of rules. This approach was often explained as a reaction to a former project, the National Clearinghouse for Geo-information, which they thought had failed due to top-down enforcement of standards. Geo-portals were to stay away from that.

Discussions always came down to technical matters—for example:

A: Technology is not the problem anymore; there are no limits, and all the essential techniques are at our disposal.
B: That's right; we should look at organizational aspects. The U.S. example of Geospatial One Stop is able to put everything they have on the Web, without restrictions.
C: But its quality is doubtful at best; they don't guarantee its accuracy. I wonder if anybody actually uses it.
A: We have to do better.
B: Just use the right standards. We have the right architecture.
A: If we stick to proven technology and standards, nothing can go wrong.
B: But what is that? Which standard is proven? Which one really works?
C: Here we go again!

In a brainstorming session in November 2005, a user-driven approach was developed: "able to find and allowed to use" (in Dutch, *Kunnen vinden en mogen gebruiken*). The approach was presented to a GI audience at the first Geo-Portals Conference in December 2005. With 50 GI professionals in attendance, the core team demonstrated that the project was user driven, but the presentation barely raised the interest of the audience. On the contrary, the presentation of the red portal Web site had an astonishing effect as it showed the project's importance and its reliance on the proper and well managed technology.

7.3.2 Middle Course: Dealing with Uncertainty

Monthly project meetings were dominated by strategic discussions. Lay users should be provided with software services for integrating, harmonizing, and presenting data; professionals should collect data themselves. This strategy was exemplified by the core team with the recurring example of a beer brewer in need of geo-information to assist in finding a location for a new brewing facility (Van de Laak 2007). The management of digital rights was regarded as an essential but separate entity of geo-portals, unrelated to technological issues.

At the end of 2006, the project team began to feel uncomfortable about the lack of guidance from SGI, which was seen as the custodian of the national geo-information infrastructure. They did not feel supported by SGI, which was reluctant to implement new ideas for NGII. SGI published an article in a leading professional magazine with the provocative title: "Where to with the Dutch Geo-information Infrastructure?" (Bregt and Meerkerk 2006). It demonstrated that SGI had no strategy and provoked discussion in the project team:

A: We are supposed to work on the Dutch NGII at geo-portals, but SGI does not help to connect to other NGI projects.
B: Are we supposed to deliver something that actually works?
C: We are working on a data viewer, but to what standards should it comply? Who is going to use it?

A: They say that a new GI-coordinating organization is in the making—yet another organization that is supposed to organize something. This does not sound like coordination to me!
D: I think that as a geo-portals team we should take a stand and do what SGI refuses: Take the lead!

After making choices on what technology and standards to apply, the geo-portals project was putting a core team in place, with representatives of three government-supported knowledge institutions and a software company to manage software development. In spring 2007, software engineers from the core team members' organizations completed a data viewer—a software device for retrieving geo-data from different sources. The data viewer was seen as an essential requirement for bringing the ultimate goal—a system of geo-portals—one step closer.

Although enthusiastic at first, the project members soon felt that the data viewer was quickly becoming outdated. They decided to enhance it with newly available techniques and developed an even more sophisticated viewer. Thus, while already having a finished and approved product ready for implementation, the development process was prolonged because new technology was explored and considered.

7.3.3 Outcome: Toward a Judgment Day

In 2007, software development went well, but the project team was becoming increasingly restless as initial goals were moving out of reach; it was felt that nobody was in charge of developing the NGII. SGI was depicted as abstract with no practical knowledge, as demonstrated by the appraisal of its promotional conference held in March 2007:

A: Real sharing of geo-information is further away than ever. The SGI conference in Rotterdam lacked any ambition. The bottom line was: "The NGII has to be developed, but let's move on as we did."
B: The "usual suspects" were doing their ritual thing.
C: It was like being in some religious rally: people celebrating and praising something of which everybody has a different image.
B: When a breakthrough is needed, nobody wants to change; we keep on doing things the way we did.
C: Everybody mentions the costs of an NGII, not the benefits.
A: If we only want an NGII for incident management and fighting terrorism, we're on the wrong track.

Project members felt that geo-portals had offered technical solutions, while management was obstructing their application. Perceptions of the role of geo-portals started to change:

> To build an infrastructure is unattainable if you just look at the budget: It was clear even before we started that it was insufficient. We had to deliver building blocks for an NGII. We are good at technology, so, when

> asked, we will handle that. But without guidance from SGI it is impossible to develop an NGII. We offer best practices and software tools. We form a community for NGII development.

The theme of the working conference in November 2007 was "Just Do It." In workshops, it was demonstrated that newly developed software applications by geo-portals were fully capable of integrating geo-data from different sources.

In search of sustainable results, project goals were redefined and the geo-portals team started to see itself as a "community of practice" of knowledge creation, which changed the atmosphere from distress to optimism to euphoria. At the closing conference in December 2008, there was confidence about project results. Software applications were presented as stepping stones in a continuous progression using newly developed techniques. Both the core team and the audience were optimistic about the future.

In interviews held after the completion of the project, technology was hailed and standards were seen as a thing of the past. Web 2.0 would make standardization redundant as it was hampering innovation in the geo-information sector.

7.4 Analysis

Using the concepts of narrative setting, space, and storyboard, we try to make sense of the three phases of geo-portals. The project itself had a clear beginning and ending; there were also some preparatory activities important for the analysis, as well as the impact of the project on the Dutch GI-sector.

Initially, the geo-portals proposal is all about developing an infrastructure serving societal needs. These needs are going to be converted into user profiles with different demand structures. As project participants become dissatisfied with the lack of guidelines for an overarching strategy, they start to develop software applications. Because they consider themselves to be at the vanguard of ever changing technology, the idea of building an infrastructure slowly fades. Instead, the goal shifts toward providing a toolbox, thus changing the image of the project as stimulating innovation.

7.4.1 A Technology-Dominated Narrative Setting

The narrative setting, dominated by rapidly developing information technology, encourages project participants to look toward the future. Geo-portals act as a means to deal collectively with the task to apply state-of-the-art technology to create newly developed software applications. Geo-portals project members, acting collectively and independently of their respective organizations, start to promote new technology, unintentionally ensuring that none of the individuals or organizations can be blamed for failure. The geo-portals project is supposed to be beneficial to the technology-dominated GI sector as a whole and is meant to supply cutting-edge technology-based methods to all.

While technology rules, geo-portals project participants have a feeling at the outset that outdated technology has been an impediment to infrastructure development and that their main objective is to put that situation to an end. Disclosure of geo-information is seen as beneficial to society as a whole, to be delivered through sophisticated GI-technology, to be mastered through the application of standards that will result in an infrastructure with a rather static form, and divided into thematically organized compartments of data, giving it a neatly arranged appearance.

As the project is progressing, technology becomes the main issue. Services have to be developed to serve lay users, and it is felt necessary to apply even more novel technology. Standards are still considered important, but now appraised as being of lesser concern. Legal aspects, although regarded as significant, are treated as a separate area with which to deal.

Toward the end of the geo-portals project, technology is seen as of utmost importance, taking the form of an unleashed phenomenon and relabeled as "innovation" in order to handle it. Innovation is thus seen as an enabler of dynamic geo-information management, without being chained by standards. However, these technological innovations are found to be obsolete before they can be used—not because they do not function properly, but rather because they are continuously superseded by even more sophisticated technology.

7.4.2 A Self-Reliant Narrative Space

At the outset, SGI as a comprehensive program is seen as an enabling organization, acting on behalf of the Dutch GI community, of which the geo-portals project is seen as only a project and subsequent to SGI, but also as beneficial to the whole sector. The geo-portals project team sees SGI as a temporary funding organization, but also as an indivisible part of the GI community and primarily involved in sustaining the geo-portals concept. This taken-for-granted relationship makes the project team cautious, responsible, and somewhat self-reflective. Thus, SGI brings the GI community together around a financing source, forcing individual organizations to cooperate in order to be eligible for funding.

After some time, the project team as a narrative space starts to develop a direct relationship with the GI community. Individual project members belong to organizations that financially support the project, but these organizations are not recognized as such within the project. As a whole, the organizations have a neutral and minor image and are all seen as equally supporting the common cause of sharing GI data. GI data users are recognized as a defined group through the user motto, even though a clear picture of these users is still missing.

The management of individual participating organizations is collectively organized into an advisory board of the SGI program. When SGI still stresses its intentions to boost innovation, geo-portals project members consider SGI to be unsupportive because it is recognized as not having a policy. SGI is seen as serving the interests of individual organizations, which do not align with the interests of geo-portals. It makes project members ready to plot their own course, which is promoting the newest trends in GI technology.

When Geonovum takes over the role of SGI, it tries to fill the gap of a lacking overall policy with an emphasis on standards; this is seen by geo-portals project members as inhibiting the possibilities created by the application of technology. Additionally, by providing insufficient funding, SGI is held responsible for not delivering geo-portals as originally planned. Realizing that the initial goals were untenable, the geo-portals team redirects its aim toward boosting innovation to facilitate the creation of an NGII. SGI was supposed to stimulate innovation in geo-information sharing, so the geo-portals project team feels quite comfortable with its new goal, knowing that its technology orientation certainly stimulates innovation.

7.4.3 Emerging Storyboards

A storyboard structures and prescribes people's actions. It may be either unconscious and tacit or prominent invoking discussion. The geo-portals project is seen as acting on behalf of the entire GI sector—detached from individual organizations and creating a stable infrastructure. With no other rules to comply with than financial and procedural ones, the geo-portals project team feels that it has to live up to its obligation, which is exploring the latest GI technology and incorporating this into a test Web site.

However, once the technology is ready to be used as a building block for GI infrastructure, efforts are put into assessing even newer technological improvements. The storyboard that can be identified here aims at the production of new technologies to be made available to the GI sector. It affects the reframing of goals, moving from the creation of a static infrastructure into making new technologies available, which is justified through the conclusion that the funds originally granted by SGI are inadequate to realize the GI infrastructure considered in the initial plan. The reframing also complies with the motto of SGI: stimulation of innovation. When the reframing takes effect, Geonovum starts to develop its own policy on the enforcement of standardization. This clear and crisp policy is fully ignored because it is seen as inappropriate to geo-portals.

The propensity to focus on technology can be conceptualized into a cyclical storyboard: Whenever new technology is tested and approved, newer technology is already virtually available to be tested and, eventually, to be confirmed as a new standard. The data convincingly demonstrate that this

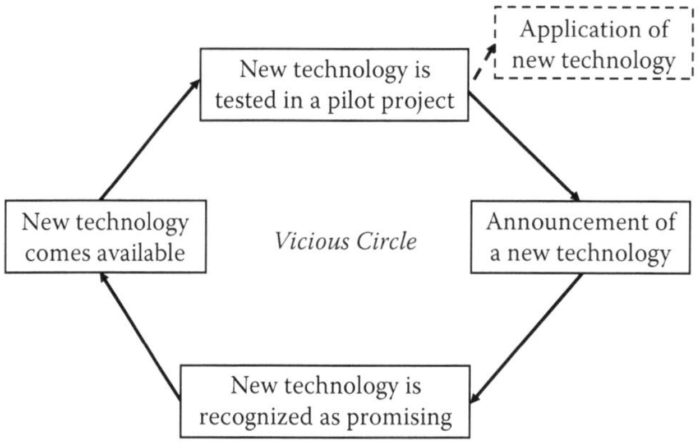

FIGURE 7.2
The storyboard of recurring action.

cycle is passed twice, following the pattern depicted in Figure 7.2. This is the storyboard of the action occurring within the project, which can also be interpreted as a vicious circle.

In a world with a pressing and increasing turnover of technological innovations, reliable infrastructures might create stability. The two competing narratives of stability and change always struggle for dominance. An infrastructure is a fixed, predictable, stable, unambiguous, and ubiquitous facility that users almost take for granted (Edwards et al. 2007). A focus on the development of a standardized infrastructure utilizes the narrative of stability, a prominent feature in the initial geo-portals project proposal (Hanseth, Monteiro, and Hatling 1996). The difficulties involved in standardization are already recognized in the project's subtitle: "liberty united." A strict regime of standardization is feared and also considered difficult to implement. Therefore, a limited, "light" version of standardization is proposed.

Throughout the project, from the initial presentation of the red portals up until the end, when the entire geo-portals endeavor was declared innovative, the emphasis was on change. The continuing process of developing software, making products already obsolete on the day of their realization, was not considered as problematic. It was seen as essential because the average GI professional seems to regard tomorrow's technology as the solution to problems encountered today.

The storyboard of innovation remains prominent. The core message of SGI is to be innovative; however, that hampers the development of an infrastructure. For this reason, the project is reframed into a knowledge-generating endeavor, driven by a storyboard of innovation. Ultimately, the GI community judges the project on its innovative qualities, demonstrated through state-of-the-art software. While this is a tangible result of 4 years of geo-portals, it is only temporary, without any reference to infrastructure.

7.5 Conclusion

Delivering infrastructure seems to involve two contradictory aspects. On the one hand, there is a narrative of change expressing the urge to work with the newest technology; on the other hand, there is a narrative of stability that sees infrastructure as predictable, stable, and thus useful (Edwards et al. 2007; Hanseth et al. 1996). These two narratives are fighting for attention.

Because the geo-portals project aimed at innovation, the narrative of change identified by the innovation storyboard was dominant. Infrastructure development rather than infrastructure building was paramount, and thus a stable, recognizable infrastructure remained absent. These contradicting narratives reflect a basic stability/change contradiction (Douglas 1986). The confrontation of two differing narratives is not uncommon and has been called the "innovation paradox," which is found in large, public sector projects where a fixed infrastructure has to be delivered in an unstable environment (Veenswijk 2006).

It has been suggested that when problems with the construction of infrastructures emerge, it is necessary to focus on project designs in the light of cultural settings (Berendse, Duijnhoven, and Veenswijk 2006; Veenswijk and Berendse 2008). However, here there is more at stake. A GI community, seemingly preoccupied with innovation, is desperately in need of a useable infrastructure. One of the project participants suggested that infrastructures are always in a process of innovation and should be regarded as "moving targets"; however, in order to be used, they need to be stable. Thus, the sector as a whole has to find equilibrium between stability and change in relation to infrastructure.

References

Barley, S. 1990. The alignment of technology and structure through roles and networks. *Administrative Science Quarterly* 35:61–103.

Bartunek, J. 2004. The importance of contradictions in social interventions. *Intervention Research* 1:103–113.

Beech, N., and C. Huxham. 2003. Cycles of identity formation in interorganizational collaborations. *International Studies of Management & Organization* 33:28–52.

Berendse, M., H. Duijnhoven, and M. Veenswijk. 2006. Editing narratives of change. Identity and legitimacy in complex innovative infrastructure organizations. *International Journal on Culture, Organization and Management* 2:255–367.

Berger, P. L., and T. Luckmann. 1967. *The social construction of reality*. Harmondsworth, England: Penguin Books.

Bergson, H. 1946. *The creative mind*. Westport, CT: Greenwood Press.

Bijker, W. 1995. *Of bicycles, bakelites, and bulbs, toward a theory of sociotechnical change.* Cambridge, MA: The MIT Press.

Boje, D. 1995. Stories of the storytelling organization: A postmodern analysis of Disney as "Tamara-land." *Academy of Management Journal* 38:997-1035.

______. 2001. *Narrative methods for organizational and communication research.* London: SAGE Publications Ltd.

Box, P., and A. Rajabifard. 2009. NGII governance: Bridging the gap between people and geospatial resources. GSDI-11. Rotterdam, The Netherlands.

Bregt, A., and J. Meerkerk. 2006. Waarheen met de nationale geo-informatie infrastructuur? *Geo-info* 7/8:296–301.

Bruner, J. 1991. The narrative construction of reality. *Critical Inquiry* 18:1–21.

Budhathoki, N., and Z. Nedović-Budić. 2007. Expanding the spatial data infrastructure knowledge base. In *Research and theory in advancing spatial data infrastructure concepts,* ed. H. Onsrud, 7–31. Redlands, CA: ESRI Press.

Burke, K. 1969. *A grammar of motives.* Berkeley: University of California Press.

Burrell, G. 1992. Back to the future: Time and organization. In *Rethinking organization, new directions in organization theory and analysis,* ed. M. Reed and M. Hughes, 165–183. London, UK: SAGE Publications Ltd.

______. 2000. Time and talk. *Organization* 7:371–372.

Crompvoets, J. 2006. National spatial data clearinghouses, worldwide development and impact. PhD dissertation. Wageningen, The Netherlands: Wageningen Universiteit.

Crompvoets, J., A. Rajabifard, B. Van Loenen, and T. Delgado Fernández, eds. 2008. *A multi-view framework to assess SDIs.* Melbourne, Australia: University of Melbourne.

Czarniawska-Joerges, B. 1998. *A narrative approach to organization studies.* Thousand Oaks, CA: SAGE Publications Inc.

De Bree, F., and A. Rajabifard. 2005. Involving users in the process of using and sharing geo-information within the context of SDI Initiatives. *GSDI-8 Conference and FIG Working Week.* Cairo, Egypt.

Douglas, M. 1986. *How institutions think.* Syracuse, NY: Syracuse University Press.

Duijnhoven, H. 2008. Tales of security practices within Spanish and Dutch railway operators: Translation, transformation or transgression? *8th International Conference on Organizational Discourse,* London, UK.

Edwards, P., S. Jackson, G. Bowker, and C. Knobel. 2007. Understanding infrastructure: Dynamics, tensions, and design. Report of a workshop on "History & Theory of Infrastructure: Lessons for New Scientific Cyberinfrastructures." September 20–October 1 2006. National Science Foundation.

Frye, N. 1957. *Anatomy of criticism: Four essays.* Princeton, NJ: Princeton University Press.

Gabriel, Y. 2000. *Storytelling in organizations, facts, fictions, and fantasies.* Oxford, UK: Oxford University Press.

Garfinkel, H. 1984. *Studies in ethnomethodology.* Cambridge, MA: Polity Press.

Gastelaars, M. 2008. Talking stuff: What do buildings tell us about an organization's state of affairs? In *Proceedings of 8th International Conference on Organizational Discourse 2008,* London, UK.

Georgiadou, Y., and F. Harvey. 2007. A bigger picture: Information systems and spatial data infrastructure research perspectives. In *Proceedings of the 10th AGILE International Conference on Geographic Information Science, the European Information Society: Leading the Way with Geoinformation,* Aalborg, Denmark.

Georgiadou, Y., and V. Homburg. 2008. The argumentative structure of spatial data infrastructure initiatives in America and Africa. In *International Federation for Information Processing,* ed. C. Avgerou, M. Smith, and P. van den Besselaar, 282:31–44. Boston, MA: Springer.

Georgiadou, Y., S. K. Puri, and S. Sahay. 2005. Towards a potential research agenda to guide the implementation of spatial data infrastructures—A case study from India. *International Journal of Geographical Information Science* 19:1113–1130.

Gergen, K. 1994. *Realities and relationships, soundings in social construction.* Cambridge, MA: Harvard University Press.

Golden-Biddle, K., and K. Locke. 1993. Appealing work: An investigation of how ethnographic texts convince. *Organization Science* 4:595–616.

Grant, D. 2003. Introduction: Struggles with organizational discourse. *Organization Studies* 25:5–13.

Grant, D., T. Keenoy, and C. Oswick. 1998. *Discourse + organization.* London: SAGE Publications Ltd.

Hanseth, O., E. Monteiro, and M. Hatling. 1996. Developing information infrastructure: The tension between standardization and flexibility. *Science, Technology & Human Values* 21:407–426.

Harré, R. 1976. *Life sentences, aspects of the social role of language.* London: John Wiley & Sons.

Harvey, F. 2001. Constructing GIS: Actor networks of collaboration. *Journal of Urban and Regional Information Systems Association* 13:29–37.

Hatch, M., and D. Yanow. 2003. Organization theory as an interpretive science. In *The Oxford handbook of organizational theory,* ed. C. Knudsen and H. Tsoukas. Oxford, UK: Oxford University Press.

Hoogerwerf, M., and B. Vermeij. 2005. Geoportal Framework version 0.2 (Geoloketten Raamwerk versie 0.2).

Koerten, H. 2007. Blazing the trail or follow the yellow brick road? On geoinformation and organizing theory. In *GI Days—Young researchers forum, 10–12 September 2007,* ed. F. Probst and C. Kessler, 85–104. Münster, Germany: Institut für Geoinformatik, Universität Münster.

Lance, K., Y. Georgiadou, and A. K. Bregt. 2009. Cross-agency coordination in the shadow of hierarchy: "Joining up" government geospatial information systems. *International Journal of Geographical Information Science* 23:249–269.

Latour, B., and S. Woolgar. 1986. *Laboratory life: The construction of scientific facts.* Princeton, NJ: Princeton University Press.

Lefebvre, H. 1991. *The production of space.* Oxford, UK: Blackwell Publishers.

Lipsky, M. 1980. *Street-level bureaucracy: Dilemmas of the individual in public services.* New York: Russell Sage Foundation.

Martin, E. 2000. Actor-networks and implementation: Examples from conservation GIS in Ecuador. *International Journal of Geographical Information Science* 14:715–738.

Masser, I. 2005. *GIS worlds: Creating spatial data infrastructures.* Redlands, CA: ESRI Press.

Nebert, D. 2004. *The SDI cookbook.* http://www.gsdi.org/pubs/cookbook/ (accessed August 5, 2010).

Obermeyer, N., and J. Pinto 2008. *Managing geographic information systems,* 2nd ed. New York: Guilford Press.

Orlikowski, W. 2000. Using technology and constituting structures: A practice lens for studying technology in organizations. *Organization Science* 11:404–428.

______. 2007. Sociomaterial practices: Exploring technology at work. *Organization Studies* 28:1435–1448.

Polkinghorne, D. 1988. *Narrative knowing and the human sciences.* New York: State University of New York Press.

Rajabifard, A., and I. Williamson. 2001. Spatial data infrastructures: Concept, SDI hierarchy and future directions. In *Proceedings of Geomatics '80,* Tehran, Iran.

RAVI. 2003. Space for geo-information BSIK knowledge project proposal. Amersfoort, The Netherlands: RAVI.

Reed, M. 2000. The limits of discourse analysis in organizational analysis. *Organization* 7:524–530.

Ricoeur, P. 1973. The model of the text: Meaningful action considered as text. *New Literary Society* 5:91–120.

Schein, E. 1996. Culture: The missing concept in organization studies. *Administrative Science Quarterly* 41:229–240.

Schmidt, A., and G. Nieuwenhuis. 2002. Geoportals "liberty united." Project proposal for Space for Geo-information. Wageningen, The Netherlands: University of Wageningen.

Schneider, B. 1987. The people make the place. *Personnel Psychology* 40:437–453.

Scott, W. R. 1995. *Institutions and organizations.* London: SAGE Publications.

Shenhav, Y., and E. Weitz. 2000. The roots of uncertainty in organization theory: A historical constructivist analysis. *Organization* 7:373–401.

Tesselaar, S., I. Sabelis, and B. Ligtvoet. 2008. Digesting stories—About the use of storytelling in a context of organizational change. *8th International Conference on Organizational Discourse 2008,* London, UK.

Van de Laak, D. 2007. DVD: Alles draait om geo. The Netherlands, Ruimte voor Geo-Informatie.

Veenswijk, M. 2006. Surviving the innovation paradox: The case of megaproject X. *Innovation Journal: The Public Sector Innovation Journal* 11: article 6, 1–14.

Veenswijk, M., and M. Berendse. 2008. Constructing new working practices through project narratives. *International Journal of Project Organization and Management* 1:65–85.

Verduijn, K. 2007. *Tales of entrepreneurship, contributions to understanding entrepreneurial life.* Amsterdam: Vrije Universiteit Amsterdam.

Warnest, M., K. McDougall, A. Rajabifard, and I. P. Williamson. 2003. Local and state based collaboration: The key to unlocking the potential of SDI. *Spatial Sciences 2003,* Canberra, Australia.

______. 2005. A collaborative approach to building national SDI in federated state systems: Case study of Australia. *GSDI-8 Conference.* Cairo, Egypt.

Watson, T. 2000. Ethnographic fiction science: Making sense of managerial work and organizational research processes with Caroline and Terry. *Organization* 7:489–510.

Weick, K. E. 1995. *Sensemaking in organizations.* London: SAGE Publications.

Wenger, E. 1998. *Communities of practice: Learning, meaning and identity.* Cambridge, UK: Cambridge University Press.

Yanow, D. 1995. Built space as story: The policy stories that buildings tell. *Policy Studies Journal* 23:407–422.

Ybema, S. 2008. Constructing collective identity: Central, distinctive and enduring characteristics. *8th International Conference on Organizational Discourse 2008,* London, UK.

Section 3

People and Practices

8

Enlisting SDI for Urban Planning in India: Local Practices in the Case of Slum Declaration

Christine Richter, Gianluca Miscione, Rahul De′, and Karin Pfeffer

CONTENTS

8.1 Introduction

Initiatives to develop spatial data infrastructures (SDIs) have been taken up since the early 1990s in many countries around the world, including India. However, the spirit of global SDI convergence that dominated the 2009 GSDI conference is more of an aim for the future than an empirically grounded phenomenon. Whether or how "the walls of SDI are coming down" (Van Loenen, Besemer, and Zevenbergen 2009, pp. 1–2) remains to be seen. It is also too soon to conclude that these initiatives automatically and unequivocally lead to wider societal benefits.

Researchers of information systems and infrastructure underline that infrastructures develop in response to different local demands. They may emerge as sociotechnical constellations in a variety of ways and often with unanticipated (positive and negative) effects (Ciborra, Braa, and Cordella 2000; Walsham, Robey, and Sahay 2007). Such variety in local place-bound circumstances and responses makes the development of large-scale systems a difficult task. In developing regions, failures of information systems outnumber successes in part due to two overlaid gaps. The first is between (Western or Western-inspired) design context and the "local actuality of the users" and the second between "hard" rational design and "soft" political and behavioral actuality (Heeks 2002). In the case of SDI, what are these design–local actuality gaps in different contexts around the world? We address this question for the case of urban India by confronting the expected role of SDI with local planning practices in a southern Indian city.

In this work we refer to "SDI development" as the overall endeavor of designing, implementing, using, and evaluating SDI. In analyzing local planning practice, we view SDI development through what Star and Ruhleder (1994) call an "infrastructural inversion … [which] de-emphasizes things or people as the only causes of change, and focuses on infrastructural relations … and reveals how choices and politics embedded in such systems become articulated components" (p. 253). Instead of predefining various components of SDI (e.g., data, networks, and copyrights) or the roles of stakeholders in SDI development (e.g., users, providers, and managers), we put the local context into the foreground—more specifically, the current dynamic relations between people, their work, and embedded information in paper and digital form.

We first provide a theoretical sketch of the role SDIs are expected to play locally. We then examine the planning context of an Indian city* and its practices in the slum declaration process through an analytical lens borrowed from Mol and Law (2002). This approach helps to explain the interplay of various actors and the multiplicity of representations in lists and drawings (of slum areas and inhabitants) created in this process. We then compare our findings with SDI expectations and discuss the implications for future SDI development at the local level. Finally, we address limitations of this study, as well as its contribution to future research.

8.2 SDI Research and Expectations

In the following section we explain the need to bring local practice to the foreground in SDI research and we sketch out the expected role of SDI locally.

* The city is one of the fieldwork locations of a 5-year Indo-European research program, which seeks to investigate the potential use of SDI by governance networks to tackle urban deprivation in five Indian cities. It is also located in a state where one of the first Indian geo-portals was launched.

8.2.1 The Missing Context of SDI Development: Local Practices

The development of spatial data infrastructure involves data, people, networks, and data access and sharing policies, as well as legal issues related to data rights, ownership, and privacy. Current approaches to SDI research seek to capture these aspects as well as the dynamics of SDI by means of increasingly elaborate sets of indicators. These are used to evaluate countries' readiness to undertake SDI (T. Delgado Fernández, M. Delgado Fernández, and Espín Andrade 2008), the national status of SDIs (Eelderink, Crompvoets, and De Man 2008; Vandenbroucke, Janssen, and Van Orshoven 2008), and SDI performance or its ability to deliver promised benefits and outputs (Giff 2008; Janssen 2008; Steudler, Rajabifard, and Williamson 2008). In these approaches, SDI is viewed as a framework that links various components at different scales—for example:

data, people, and networks (Rajabifard 2008, p. 13)

policy, data, people, and networks (Steudler et al. 2008, p. 205)

data users and providers (Janssen 2008, p. 262)

different administrative levels of government and nongovernmental organizations (NGOs) and sectors (Masser, Rajabifard, and Williamson 2007, p. 10).

These approaches are rather data centric. Social dimensions, especially, are considered only as factors influencing data dissemination and use. People and their relation to information are usually summarized as "human resources," "people component," "stakeholders," or simply "people." Keeping (potential) stakeholders and their practices in the background is problematic because, as Nedović-Budić, Pinto, and Raj Budhathoki (2008) note, "[u]ser concerns and the level of use of an innovation … do not exist in a vacuum" (p. 278).

Some studies offer a deeper understanding of the relations between people and SDI. For instance, they examine the roles and perceptions of different stakeholders in the Indian National SDI development (Georgiadou, Puri, and Sahay 2005; Puri 2006; Puri, Sahay, and Georgiadou 2007), personal characteristics of key individuals driving SDI in the Minnesota SDI (United States; Craig 2005), or how SDI practitioners themselves evaluate their development efforts (Lance, Georgiadou, and Bregt 2006; Lance 2008).

Data sharing studies, particularly, investigate the relation between people's practices and information. Elwood's (2007) research in Chicago, Illinois, shows how land use data sets from the municipality are of little use to grassroots organizations because of differences in the epistemologies that stem from the NGO's objectives and day-to-day practice. Schuurman and Leszczynski (2006) and Schuurman (2005) explain how differences in land use classifications of two county government databases are embedded in different planning practices and rationales, making the data ontologically incomparable despite achieving technical interoperability.

A few studies focus on the relations between people's practices and newly introduced technology within a specific local context (Harvey, 2006, and Silva, 2007, for digital cadastre and land administration systems, respectively). In India, Puri et al. (2007) emphasize the importance of locally relevant hardware and software and the need to recognize multiple stakeholders and exercise social sensitivity for the success of several local technological initiatives. De' (2006, 2008, 2009) investigates the relationship between caste structure and e-governance projects in India and finds that the systems favor dominant castes in terms of design and benefits. These researchers explicitly address existing relations between people and land as they have developed historically in a specific context. An understanding of these historical contingencies allows researchers to investigate how system implementation shapes and is shaped by existing practices (in land administration, for example).

The paucity in SDI research addressing the relations between SDI and people's practices within specific local contexts is problematic in three ways. First, there is evidence that existing practices play an important role in the trajectory SDI development takes. For instance, Davis and Fonseca (2006) note that the "success [of SDI] has primarily been a result of reaching wide agreements on principles and practices, always guided by real needs and applications" (p. 288); Harvey and Tulloch (2006) emphasize that data sharing does not happen in and for itself, but is always part of other activities. Neglecting the local context of SDI development also makes it tempting to apply standard indicators of success and failure across diverse settings and situations globally, while various settings may pose different challenges and opportunities.

Second, the knowledge of local context can facilitate the identification of opportunities for local institutionalization without having to assume that geographic information types and uses play the same role everywhere. For example, in our research we find that local address systems rely heavily on landmark descriptions and that photographs of construction sites play a more important role in planning and budget meetings than thematic or topographic maps.

Third, an understanding of existing local practice is necessary to assess how far SDI development can address wider social goals, like management of disasters, resources, and environment; alleviation of poverty; or access to land, housing, and physical infrastructure.

The need for more sensitivity to the context has been noted elsewhere. Nedović-Budić et al. (2008) suggest that evaluations of SDI ought to be carried out in conjunction with contextual factors and determinants of outcomes. Georgiadou and Stoter (2008) explain the need to study the dynamic relations between social context and geoICT in use through longitudinal, interpretive and in-depth case studies. For information and communication technology (ICT) in general, Prakash and De' (2007) emphasize how the context influences the value of ICT to socioeconomic development outcomes. Although her recent book is titled "e-Governance for Development—A Focus on Rural India" (2009),

Madon chooses to put "greater emphasis on governance and development, rather than on the 'e' [because of] the important role played by local governance structures in addressing the development needs of the community" (p. 6).

In places where SDI is only emerging or where state and national SDI efforts have not touched the ground yet, various fields of practice (for example, resource management, administration, or planning) can provide contextual boundaries for data collection and analysis. We focus on people's practices in local urban planning. In order to analyze these practices in a way relevant for local SDI development, we first sketch out the expected role for SDI locally. To do this, we have to rely on Indian SDI strategy documents for the national and state levels.

8.2.2 SDI's Expected Role as Ordering Mechanism

The aim of this section is to provide a theoretical sketch of the expected role of SDI locally. We review the 2001 Indian NSDI Action Strategy Plan (GOI-DST 2001) and the 2008 Progress Report for the state SDI.* We draw mainly on a section in the NSDI document called "Spatial Information—Indian Perspective" because this section provides insights about the type of geographic information, its expected use, and locally involved actors.

The state SDI was launched at the NSDI-11 conference in Pune in 2009 as the second state geo-portal in India after the New Delhi geo-portal. Sanctioned in April 2007, it is a joint initiative of the Natural Resources Data Management System (NRDMS), the national Department of Science and Technology (GOI-DST), and the state government. The state geo-portal seeks to support interagency data sharing through a common platform, identify the needs of various government programs, and facilitate decision making and local-level planning. In the following we make these expectations for the local role of SDI more specific. The state SDI is expected to support the collection and integration of geographic information about various aspects of the world. In the Indian documents, there is strong emphasis on the use of geographic information systems technology allowing users to "crunch together" data and process maps to provide spatial visualization of information (GOI-DST 2001, p. 1.2).

The outcome of such assemblage is to be "used for a wide range of applications—natural resources management, wasteland development, watershed development, urban management, coastal management, utilities management, infrastructure development, business development, etc." (GOI-DST 2001, p. 3.1).

* To retain anonymity of informants, we do not cite exact names for this Indian state and refer to the state as "state" in document names in the text. The Action Strategy Plan was chosen over a more recent account of the Indian NSDI efforts because the former outlines expectations and a course of action. The progress report for the state geo-portal is relevant because the city we study is located in this state and the document is the most coherent source available at the time of writing.

Outcomes of information integration must be considered representations of the world—for example, representative of all areas that are most prone to flood in a given country or inclusive of all corridors with low and medium traffic volumes for a given urban area. Although GIS offers the opportunity for users of information to "become mappers" and "many possible mappings could be made," mapping "still depends upon a representational view of the world" (Kitchin, Perkins, and Dodge 2009, pp. 7–8). The third message in the documents relates to the actors expected to assemble geographic information locally. For the NSDI, the government is viewed as the major enabler (GOI 2001, pp. 5.1–5.2).

According to the progress report for the state geo-portal (State Council for Science and Technology 2008), assemblers of information are the offices of public administration and associated organizations (the document also lists state-level boards). Specifically, for local planning in cities, the assembler is expected to be the urban local body: "The process and practice of infrastructure management are being decentralized to the PRI's/Urban Local Bodies (ULBs) in order to make the related strategies area-specific and responsive to the local aspirations and needs" (State Council for Science and Technology 2008).

We can sum up three key expectations for the role of the Indian SDI locally:

- To collect and combine geographic information about the world (e.g., city, region, and state)
- To produce outputs from data combination in GIS—for example, as thematic maps—that are representations of the world (domain or application specific) to be used in management, planning, and decision making
- To connect to the local level, where mainly formal organizations of the public administration, especially the urban local bodies, are responsible for collecting and combining geographic information for use in urban planning (at least initially)

The underlying rationale of these three key expectations can be conceptualized through Mol and Law's (2002) concept of order. Orders in this sense "do not simply expel the complex and chaotic. In addition, they insist that what belongs to them is drawn together and properly assembled. No element may hold back, and what is inside must be named, accorded a place" (pp. 13–14). Information is not randomly grouped together. In ordering, the aim is to assemble elements to fit into a larger scheme. The data in GIS tables and remotely sensed images and on maps are then viewed as representations of some aspect of the world (flood-prone areas, low-traffic streets, etc.). Mol and Law (2002) identify three modes of representation from which order emerges. SDI's expected role as ordering mechanism involves two of these modes.

One is the classificatory system. A table or database that classifies "presupposes a single and conformable world ... and makes big cages that are then subdivided into smaller ones" (Mol and Law 2002, p. 14). An example of a classificatory system as mode of representation is land use classification. The aim is to include every possible land use, and this is done through hierarchical nesting of land use categories. The first level may be residential, industrial, commercial, and open space. The second level is low- and high-density residential areas, light and heavy industry, and so forth. The classification insists on properly assembling these elements and being all inclusive and hence representative of the world with respect to land use.

Maps are the second mode of representation relevant for SDI. They are not necessarily classifications, but rather "draw surfaces that contain details," sites, and their attributes, "that are related in an accountable manner" (Mol and Law 2002, p. 16).

The SDI's expected role locally can be conceptualized as an ordering mechanism. The assemblage of geographic information through GIS in public administration would draw on two modes of representation: classificatory system and maps.

The goal of gaining an all-inclusive overview does not mean that geographic information assemblage is expected to include and represent all of reality. Instead, "order" refers to an underlying rationale, where information is assembled in such a way that it is all inclusive with respect to certain application domains (e.g., land use planning) and where outcomes of assemblage are representative for this specific aspect of the world (e.g., land use).

We will return to the implications of conceptualizing the role of SDI as ordering mechanism later in the chapter. In the following section we describe the research methodology and then analyze existing local planning practices.

8.3 Methodology

8.3.1 Empirical Case

The context for our study of local practices is the realm of urban planning. The empirical case concerns the identification and declaration of slum areas. Slum declaration has had direct implications for the spatial allocation of funds under poverty and housing schemes since the 1970s. It is part of the larger process of slum improvement and rehabilitation. Planning here is not the practice of expert planners or the department of planning. Instead, it involves the work of various municipal and state departments, parastatal boards, and nongovernmental actors.

At the time of research, Mugdali was part of a major national urban renewal mission initiated in 2005 and scheduled to run for 7 years. Slum rehabilitation forms the focus of the second of two subprograms in the mission and entails provision of housing and physical infrastructure specifically for slum areas. The listing of declared and undeclared slums plays a central role in these activities because rehabilitation relies on the identification and declaration of areas to be rehabilitated. "Undeclared" refers to areas that are known to be slums, but are not officially declared. "Declared" means that the area is officially declared as a slum. Per mandate, the implementing agency for rehabilitation* work is the State Slum Clearance Board, established in 1975 through the State Slum Areas Act of 1973.†

8.3.2 Research Approach

Our research approach is exploratory and interpretive (Walsham 2006) because the empirical field is historically, culturally, and linguistically different from the geographical locations where the notion of SDI was first conceived—namely, in North America (Homburg and Georgiadou 2009). The time frame of this study allows us to spend extended periods in India. We are able to gather descriptions from different perspectives and go beyond what people say by observing what they do (Silverman 1998). We generate questions in an ongoing manner and follow new insights in order to build explanations gradually. We draw on 7 months‡ of ethnographic fieldwork in Mugdali, in southern India. The empirical findings presented here are derived from three main data sources:

- Government documents, especially the 1973 Slum Clearance Act§
- Semistructured interviews and informal conversations with officials and staff at the local slum office and municipal office (urban local body), ward councilors (politicians), and others
- Notes from field observations (Figure 8.1)

The first author spent, on average, 5 hours per week in the slum office, an adjacent rehabilitated slum, and a small, convenient shop in front of the slum office to gain insights into the work taking place at the office and into

* The language with respect to slum "rehabilitation" has changed during the past 40 years, depending on policy goals, from "clearance" to "improvement" and currently "rehabilitation." The latest terminology seeks to reflect policy and funding preferences for in situ development and provision of social services instead of area clearance and relocation.

† What we call the "slum office" is a divisional office of the State Slum Board. The office is responsible for slums in three districts, including Mugdali.

‡ Split in one 4-month and one 3-month period.

§ This act outlines procedures for declaration and definition criteria. It is often cited by slum officials and was used for analysis in conjunction with procedure descriptions in interviews.

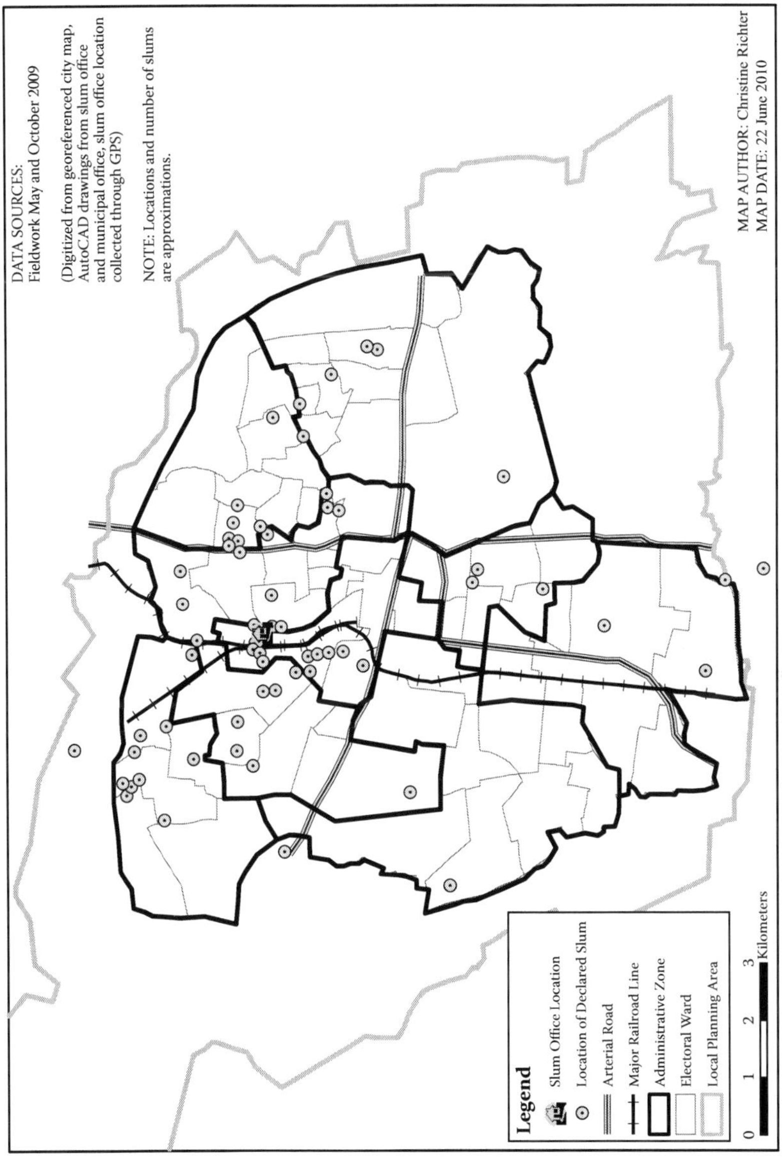

FIGURE 8.1
Location of slum office and declared slums (approximate location and number).

its connections to other actors. The research team shadowed slum office surveyors during their socioeconomic survey in one declared slum and visited 14 declared slums, two migrant camps, and one undeclared slum area repeatedly for transect walks, mapping, and interviews with dwellers,* leaders, associations, and women self-help groups. Interviews were also conducted with representatives from two NGOs active in slum organizing and improvement and one NGO funding organization.† During the first 3 months, the researchers attended weekly review meetings for the urban renewal program at the municipal office.

In addition to field notes, a field diary and interpretations/hypotheses were kept separately during fieldwork in order to ensure reliability of analysis. To exclude empirical data unrelated to slum rehabilitation activities, the analysis was performed only on dates from the field notes in which "slum" was mentioned.

For analysis, we first wrote three detailed descriptions of the declaration process from different perspectives: the process as described by slum officials and Slum Clearance Act; detailed summaries of our observations, accompanied by photographs; and a description based on information artifacts created in the process (e.g., list of declared slums, list of beneficiaries, and boundary drawings).

8.3.3 Analytical Lens

Based on these descriptions, we categorized practices according to Mol and Law's (2002) concept of lists that classify (based on classificatory system) and lists that do not classify. If we read a list that is based on some kind of classificatory system, we can understand an underlying set of criteria. It may order the world by size, by genetic similarity, by elevation and slope, etc. At the same time, such a list seeks to represent aspects of the world drawing on classificatory systems. In the following, we call the practices that bring about such lists "classificatory listings."

But a list does not have to classify and does not have to seek to impose a single order onto the world. Here, Mol and Law give the example of a Chinese encyclopedia‡ that divides animals into categories like "stray dogs" and "drawn with a very fine camelhair brush" (2002, p. 14). Such a list does not classify, at least not in any recognizable way. It only groups together without relying on any single logic of ordering. According to Mol and Law, "Items in [such] a list are not necessarily responses to the same questions but

* We use the word "dwellers" because it is locally used to refer to people who live in (declared and undeclared) slum areas (in English conversation).

† One interview was conducted in May 2009 with one representative from each organization in their offices in Mugdali. One of the interviewees invited us for a site visit to a slum recently included in the NGO's youth program. We interviewed the representative from the funding organization twice over tea in a roadside hotel.

‡ Foucault, in "The Order of Things," borrows this example from Borges.

may hang together in other ways, for instance socially, because [such] a list may be the result of the work of different people who have each added something to it" (p. 14). That is, "they assemble elements that do not necessarily fit together into some larger scheme [and] they make no claims to inclusiveness" (p. 7). In the following discussion, we call the practices that bring about such lists "nonclassificatory listings."

We use the term "listings" in order to make explicit what people are actively doing (listing). Listings are the practices through which lists and maps of slums and inhabitants are created and used. The dichotomy of classificatory/nonclassificatory serves as a sense-making tool to analyze the practices in which information artifacts are embedded.

8.4 Findings

We describe the two listings separately and explain how far they are classificatory and nonclassificatory. However, the process of slum declaration must be understood as the interplay of both listings and information artifacts must be viewed as outcomes of this interplay.

8.4.1 Classificatory Listings

One set of practices is driven by public administration procedure and staff. It follows legal and monitoring requirements, which also determine the types of information artifacts required for official declaration.

The slum office learns about an area (undeclared, but known) through NGOs and slum dweller associations, who visit the slum office or ULB. An NGO might also provide this information to the deputy (district) commissioner (DC), who then forwards it to the slum office. Surveyors visit these sites to prepare information required for declaration—namely, a boundary drawing of the area (by hand and measuring tape) and a beneficiary list with socioeconomic data (SES survey form) for each household that is either signed or fingerprinted by the household head. Slum officials check the area against legal requirements for slum declaration, where:

> ... the Government is satisfied that,
> (a) any area is or is likely to be a source of danger to health, safety or convenience of the public of that area or of its neighborhood, by reason of the area being low-lying, insanitary, squalid, overcrowded or otherwise [*sic*]; or
> (b) the buildings in any area, used or intended to be used for human habitation are,
> (i) in any respects, unfit for human habitation; or

(ii) by reason of dilapidation, overcrowding, faulty arrangement and design of such buildings, narrowness or faulty arrangement of streets, lack of ventilation, light or sanitation facilities, or any combination of these factors, detrimental to safety, health or morals,

it may, by notification, declare such area to be a slum area. (State Slum Areas [Improvement and Clearance] Act 1973, Chapter 2)

In interviews, the term "unfit for human habitation" is often expressed as "the houses are not pucca." "Pucca" roughly translates into "solid/permanent." In this case, "not pucca" refers to various physical housing characteristics, such as lack of ventilation, thatched roofs instead of metal roofs, and dirt floors. Slum officials also cite 10–15 as the minimum number of dwelling units to constitute a slum area. In addition, the surveyors check (and this can take several months) voters' lists or ration cards to evaluate length of stay of a household for eligibility. To be included in the beneficiaries' list, legal requirements must be met according to the 2004 State Slum Act Amendment Rules.

The list of beneficiaries with socioeconomic data and boundary drawings is reviewed by the assistant executive engineer (AEE) in the slum office and submitted to the ULB to receive no objection certificates as well as to the state level slum board office for approval. From there it goes to the DC, who issues a declaration by publishing the area's name in a gazetteer.

These practices in response to procedural and legal requirements are driven by the goal to order through classification. The underlying rationale has at its core three "cages" (to use Mol and Law's terminology): from city to area (based on Slum Act criteria) to individual households (based on Slum Act Amendment criteria). The objective is to identify and declare all areas and people that match these criteria—that is, to be all inclusive and to create a list representing all slum areas, dwellers, and their characteristics in Mugdali per established criteria.

Although the slum office forms a nodal point for these activities, it also involves the board's state office, several consultants, the ULB, the DC office, and, especially during land disputes, the urban development agency and courts. In other words, these practices are driven mainly by procedure and staff of the public administration. The content of information artifacts, however, varies and can be highly contested, as we witnessed in review meetings. There are, in fact, multiple representations, and no list can claim an all-inclusive overview of slums in Mugdali. This can be explained by practices beyond those driven by procedure.

8.4.2 Nonclassificatory Listings

Information artifacts are at the same time an outcome of practices of people outside public administration and the relations between them, public administration, and the urban environment, like land and temples. These practices are not driven by an underlying aim to order (through classification). Instead, "if someone comes along with something to add to the list, something that emerges as important, it may indeed be added to it" (Mol and Law 2002, p. 14). There is no one entity or criterion that decides what is important to be added to the list. New criteria emerge as important in these practices depending on changing sociopolitical situations.

The most frequent visitors at the office are leaders of the umbrella organization of slum dwellers, who negotiate throughout the process of declaration and lobby for inclusion of new areas and individual beneficiaries on lists. Leaders of individual areas and their respective associations also come directly to the office to "give presentations about their case," as an official said. Slum dwellers may approach the office with their own boundary drawings and lists of beneficiaries. Throughout the week, groups of people wait at the office to meet officials and become listed as beneficiaries.

During socioeconomic surveys in slums, officials not only write lists of houses and families, but residents also keep lists and provide this information to officials. In these cases, the dwellers (through associations, usually) directly add or seek to add to various official lists of known slums, of declared slums in the longer term, and of beneficiaries. Now the cages of classification (city–area–household) become blurred. The people and areas that are supposed to be classified based on a set of criteria participate in the creation and changing of lists, but not in an ordered fashion. The aim here is not to create an all-inclusive representative list. Practice is driven by varying objectives with respect to specific areas, groups of people, and individuals.

What these objectives are and how they can be pursued is influenced by emerging associations, knowledge of opportunities, and an area's distance to the slum office. As one of the shop owners in front of the office explained to us, it is easier to come to the office regularly to "put pressure on officials" (to move forward the process of declaration and subsequent rehabilitation) if one lives closer because of travel cost and time. "That is why you see many declared slums around the office."

What happens around the slum and other public administration offices is linked to the less visible backstage work. The AEE attributes the increase in slum declarations over the years to the work of NGOs, which raise awareness among Mugdali's urban poor to access various national and international program funds and services. Members of NGOs also emphasize that money management and substance abuse are of special concern in slum areas—more so than levels of income. There are, then, socioeconomic criteria other than the physical criteria stipulated in legislation that carry relevance and influence which areas are brought to the attention of public administration and become listed.

When we enter backstage politics, the listings become difficult to trace. Here, "things hang together" in less predictable ways and emerge from different social relations. During a site visit with one of the site engineers to a recently declared slum, we asked how the slum office came to know about this area. We learned that a member of the legislative assembly (MLA), a state-level politician, approached the office on behalf of the people. MLAs also approach state-level bureaucrats with the aim of listing party workers as beneficiaries if they live in slums. In exchange, the politician may influence the transfer of bureaucrats between public administration posts. Local-level politicians (for example, ward councilors) also enter into sociopolitical relations with slum dwellers, leaders, and public administration in return for votes. Text Box 8.1 shows excerpts from an interview with a local Congress party leader and ex-ward councilor that sheds light onto these relations from her perspective.

Whether an area becomes included in the list of declared slums also depends on land ownership and rights. It is more difficult if the slum is located on privately owned land because, in this case, people have to be relocated and may oppose declaration because it entails a move away from jobs and relatives.

TEXT BOX 8.1: INSIGHTS INTO THE RELATIONS BETWEEN POLITICIANS, SLUMS, AND PUBLIC ADMINISTRATION FROM THE PERSPECTIVE OF A LOCAL POLITICAL LEADER

Excerpts from an Interview with a Local Party Leader

(A) Why one slum was declared (and infrastructure provided) and the other not, she explains:

> Here in [slum A] the MLC [Member of Legislative Council] M.G. was there. He constructed them houses from the government fund... They have nice facilities like water, light, and all underground facility. Everything is ok for them. But these people [referring to a different undeclared slum], they cannot sleep in the night; snakes and scorpions will come ... Rainwater passes through their house. I don't know how they live.

(B) Intermediaries may also be politicians running for office at ward level. In many cases, politicians promise dwellers declaration of their area and/or provision of services in order to capture the votes of area residents. Slum dwellers may approach the politician according to the interviewee:

> How do I know about the people? When elections come, they come to our door ... They will know the candidate, who is the candidate for my area, so they will come to our door [and they say:] "You come and see my place. We need this much of work. And we are so many there, you have to give money." That is how they demand. Most slum people will vote; 90% of the people [from a slum] will come and vote. But they make demands [in return].

(C) Negotiations over money, votes, and services take place between politicians, slum leaders and public administration:

Researcher: Do you know how one person in a slum becomes the "senior"?

Politician: Because they have stayed there the longest. And then they start collecting [a type of rent from other dwellers].

Researcher: Do you think some corporators [ward councilors] support these senior people in the slum?

Politician: For the vote purpose, but they have to be silent.

Researcher: The longer they live there, the more senior they are and then they start collecting?

Politician: He [senior] is the boss now and he will tell every person whom to vote for ... He comes for everything. Like how much money has to be given per vote, what all the things are that have to be provided in return, what facilities. He will also talk to the government about that. All the others [slum dwellers] are under his control.

Researcher: He acts a little bit like an elected person then, but he is not elected?

Politician: Yes, yes.

Researcher: He is only senior?

Politician: Yes. But they also elect within their slum. If there are two persons competing, they will hold an election.

Length of stay directly influences whether residents are included in beneficiary lists, but this may in turn be an outcome of the interplay between political relations, relations to land, and socioreligious practices (as the example of temple construction shows). To interrupt a direct relation between voters (slum dwellers) and local leaders or politicians while securing the area as a vote bank, the politicians may introduce a third figure—namely, a god or goddess—through temple construction (Beck 1976; Brouwer 2004, 2007). The temple in turn makes it possible for people to stay on the land for extended periods of time because it is complicated to shift after construction.

Compared to the more procedure-driven classificatory listings, the practices that we describe here as nonclassificatory listings are more situational and ad hoc in nature, with different criteria and actors emerging through time. The nonclassificatory listings explain why it is difficult to recognize a single underlying ordering rationale that determines what is included in the lists. They explain the variability in content and also contesting versions of lists of slums in the city. Nonclassificatory listings outside public administration merge into the classificatory listings (and vice versa), forming a dynamic dialogue from which lists emerge that cannot completely fulfill their promise of comprehensively representing the world (all slums and characteristics in Mugdali). The intertwining of these two "listings" is reflected in Figure 8.2. The photo illustrates the attempts on the part of the public administration to represent the same door once and for all on the list (the survey form).

FIGURE 8.2
Repeated survey numbers on door in declared slum. (Photo: C. Richter, May 22, 2009.)

We are left with three numbers on one door and one door being recorded at least three times. This becomes plausible and explainable only when we take the interplay of both listings into consideration—an interplay characterized by shifting sociopolitical alliances and relations between people, land, and built environment.

8.5 Implications of Findings for Local SDI Development

Part of the process of slum declaration is driven by public administration procedure and staff. Information assemblage in these classificatory listings follows a rationale of order through modes of representation—specifically, classificatory systems and mapping (lists of slums and beneficiaries, boundary drawings). The expected role of SDI as ordering mechanism fits these practices. It would serve public administration in collecting and combining information in a comprehensive manner according to specified criteria and definitions of areas and people, and in turn to create representations of slums in databases and on maps.

However, actual practices in the process follow procedure (and its rationale) only partially. Classificatory and nonclassificatory listings intermingle; the assemblage of information does not rest solely in the hands of public administration. Slum dwellers, NGOs, and other organizations are not silent

TABLE 8.1

Expected Role of SDI and Local Context

Expected role of SDI	Local context
To collect and combine geographic *information about the world* (city, region, state, domain specific).	There is *no clear-cut separation between geographic information about the world* (number of dwellers per area, slum boundaries), the world, and the activities of combining information.
To produce outputs from data combination in GIS—for example, thematic maps, which *represent* aspects of the world and provide a basis for management, planning, and decision making.	Lists of slums/beneficiaries are *not a 1:1 representation* of slums/beneficiaries in a city, because lists change, are contested, and (re-) created through the interplay of classificatory and nonclassificatory listings.
To connect to the local level, where mainly formal organizations of the *public administration*, especially the urban local bodies, are responsible for collecting and combining geographic information for use in urban planning (at least initially).	There are *others who collect and combine* information, including dwellers and NGOs, and the associations they form with politicians and urban environment. Assemblage is not contained within formal organizational boundaries.

and passive data sources. Through nonclassificatory listings, they engage in the creation of lists, add to tables, and change boundaries. From the point of view of SDI as an ordering mechanism, these nonclassificatory listings pose an obstacle to an all-inclusive and representative view of slums and inhabitants in Mugdali on a final table or map.

Presently, the interplay of classificatory and nonclassificatory listings creates multiple and competing versions of slum areas and characteristics, preventing any one representation from claiming dominance over an extended period of time. Specifically, for the case of slum declaration and related planning of poverty alleviation and housing programs, SDI expectations also face the problem of shifting control over the assemblage of information *within* public administration—namely, from the slum board to the ULB. We have summarized the reflection between expected SDI role and local context in Table 8.1.

SDI development at the local level cannot be reduced to issues of ICT capacity and the development of technical standards and data sets. Instead, it may affect and require changes in planning practices, underlying rationales, and potential stakeholder roles. We discuss possible implications for SDI implementation: one in its (expected) administrative ordering role and one in an alternative, more open and flexible approach.

Local SDI development that would follow the outlined expectations for SDI as an ordering mechanism favors and encourages the practices which we identified as classificatory listings through GIS mapping. These practices are viewed separately from nonclassificatory listings, and the latter are excluded from the assemblage of information. Information gathering and other related activities are firmly centered within public administration—for example, a municipal GIS department. The role of people and their associations with each other and with the urban environment (land, temples) are removed from the realm of data assemblage. Public administration offices—specifically, the ULB office—are turned into "centers of calculation" where "observations are accumulated, synthesized, and analyzed" (Kitchin et al. 2009, p. 16) with the aid of GIS technology. If classificatory listings gain dominance, so would specific representations of the city, its slums, and residents. Those representations would be based on official government procedures and views as opposed to representations that encompass multiple perspectives and versions of reality.

A push toward classificatory rationale and control vested more strongly in the hands of public administration may have positive or negative consequences. On the one hand, the disregard for direct relationships between dwellers and street-level bureaucrats can lead to a loss of venues for negotiation. Street-level staff's and citizen organizations' intimate knowledge of the dynamic sociospatial and political relations is not included or at best moved into the background and "rigidized" in GIS databases. On the other hand, an expanded mode of classification may include areas and people who are currently ignored because they lack the opportunity to participate in either classificatory or nonclassificatory listings (e.g., migrants who do not speak the local language or who cannot access associations due to stigmatization).

Alternatively, the expected role of SDI could be more broadly conceived or changed during development through a more flexible approach. With the more open approach to SDI development, types of geographic information (e.g., dwellers per area), roles of stakeholders (NGOs and politicians as data providers), and identity of assemblers (ULB) are not predefined firmly. Instead, development would follow an "infrastructuring" approach (Pipek and Wulf 2008). In infrastructuring design, implementation, use, and evaluation are not distinct steps in a linear process and the roles of designers, implementers, and users are not fully separated. Instead, boundaries between the steps and roles in SDI development become blurred. The system is designed through use; in turn, use designs the system.

Such an approach would be neither purely "top down" nor "bottom up." The decisions of expert designers about the roles of various SDI stakeholders are replaced by opportunities for actors to shape* their own roles vis-à-vis geographic information during development. What emerges as SDI locally may be quite different from northern or Indian national and state notions and expectations. Such an approach would also require more flexible evaluation of SDI that allows assessment criteria and indicators of success or failure to emerge during infrastructuring from the specific context and based

* The term "shape" (from the shaping of technology literature) here reflects the contingent, emergent, reactive nature of the activity rather than forethought and planning as suggested by the word "design."

on wider societal goals in this context. The impacts of SDI development on planning practices and changes in stakeholder roles may receive more consideration under the open SDI scenario.

8.6 Research Limitations and Further Research

In this chapter we focused solely on the process of slum declaration. Further research is needed to investigate our findings with respect to the wider realm of urban planning and with respect to explicit local geo-ICT development efforts. Through our analysis of practices, we shed light onto the relations between people, land, environment, and information artifacts. Contesting representations of the world cannot be explained quickly by factors such as lack of comprehensive data sets or irregular table updates.

However, this study is weak with respect to Klein and Myer's (1999) fourth principle for evaluation of interpretive research: the application of findings "to more general concepts (theory) that describe the nature of human understanding and social action" (p. 72). In other words, future research needs to explain why these two rationales intermingle. In addition, the theoretical sketch of SDI's expected role locally is based on key messages from two Indian SDI documents only. Future research needs to investigate further the expectations embedded in the development and discourse of national and state SDI.

Our main contribution lies in the identification of a boundary never tackled in SDI development literature—namely, that between two rationales that mix in practice. SDI research so far has mostly addressed the objective of seamless integration of geographic information (socially and technically) and information systems vertically and horizontally. From this point of view, SDI research tends to regard heterogeneity of information (e.g., lists) as indicative of clashing classifications, which in turn are perceived to arise from different organizational mandates and procedures. For example, an organization with the objective of managing land slide risk may classify land cover differently than an organization concerned with the appropriate distribution of urban land uses. In support of comprehensive and rationalist planning, SDI would require the standardization and/or interoperability between classifications through technical solutions and through interorganizational coordination and agreements.

However, our analysis shows that information creation and use fall along a boundary of practices where one is driven by a classificatory rationale, but the other is not. The latter does not seek to order through representation according to specific definitions or criteria, but rather is driven by changing and situation-dependent objectives and alliances. It functions in part on the basis of inconsistency and heterogeneity. At the same time, the boundary between these two rationales does not necessarily align with any formal organizational boundary, nor do the two listings coincide with a state–citizen dichotomy. A public administrator may follow procedure today and follow political relations tomorrow, depending on objectives and strategies.

In conclusion, future SDI research should not only search for solutions to standardize existing information or seek to explain different classifications and definitions, but also explore the rationales and objectives of people's practices that do not easily align with organizational mandates, procedures, or even formal organizational boundaries. Questions arising for such research may include:

Do people who adapt or drive SDI initiatives rely on the rationale of classification (in discourse and practice), on nonclassificatory rationales and sociopolitical alliances, or a mix of both?

What does that mean for the two-directional shaping between SDI development and urban planning practices?

Will practice become more classificatory or will SDI become less classificatory—for example, serving as a discursive device only?

The aim of such research is to trace the incorporation of new technology into the realm of wider practices to learn and to explain how these are shaped by and are shaping technology and what the consequences might be.

Acknowledgments

The authors would like to thank the staff of the municipal and divisional offices in Mugdali for their assistance in data collection as well as the Dutch Science Foundation for funding this research in the framework of the NWO-Wotro-Integrated Program, "Using Spatial Information Infrastructure in Urban Governance Networks: Reducing Urban Deprivations in Indian Cities."

References

Beck, B. E. F. 1976. The symbolic merger of body, space and cosmos in Hindu Tamil Nadu. *Contributions to Indian Sociology* 10:213–243.

Brouwer, J. 2004. An exploration of indigenous knowledge for an anthropology of India. In *Studies in Indian Anthropology*, ed. P. K. Misra, 30–46. Jaipur: Rawat Publications.

———. 2007. Culture and contrasting views on the individual, autonomy and mortality with special reference to India. In *Autonomy beyond Kant and hermeneutics,* ed. P. Banerjee and S. Kumar Das, 3–36. London: Anthem Press.

Ciborra, C., K. Braa, and A. Cordella. 2000. *From control to drift: The dynamics of corporate information infrastructure.* Oxford, England: Oxford University Press.

Craig, W. 2005. White knights of spatial data infrastructure: The role and motivation of key individuals. *URISA Journal* 16:5–13.

Davis, C. A., Jr., and F. Fonseca. 2006. Considerations from the development of a local spatial data infrastructure. *Information Technology for Development* 12:273–290.

De', R. 2006. Evaluation of e-government systems: Project assessment vs. development assessment. *Lecture Notes in Computer Science* 4084:317–328.

———. 2008. Control, de-politicization and the e-state. *Lecture Notes in Computer Science* 5184:61–72.

———. 2009. Caste structures and e-governance in a developing country. *Lecture Notes in Computer Science* 5693:40–53.

Delgado Fernández, T., M. Delgado Fernández, and R. Espín Andrade. 2008. The spatial data infrastructure readiness model and its worldwide application. In *A multi-view framework to assess spatial data infrastructures,* ed. J. Crompvoets, A. Rajabifard, B. Van Loenen, and T. Delgado Fernández, 117–134. Wageningen, The Netherlands: Space for Geo-Information (RGI).

Eelderink, L., J. Crompvoets, and W. H. E. De Man. 2008. Towards key variables to assess national spatial data infrastructures (NSDIs) in developing countries. In *A multi-view framework to assess spatial data infrastructures,* ed. J. Crompvoets, A. Rajabifard, B. Van Loenen, and T. Delgado Fernández, 307–326. Wageningen, The Netherlands: Space for Geo-Information (RGI).

Elwood, S. 2007. Grassroots groups as stakeholders in spatial data infrastructures: Challenges and opportunities for local data development and sharing. *International Journal of Geographical Information Science* 22:71–90.

Georgiadou, Y., S. K. Puri, and S. Sahay. 2005. Towards a potential research agenda to guide the implementation of spatial data infrastructures—A case study from India. *International Journal of Geographical Information Science* 19:1113–1130.

Georgiadou, Y., and J. E. Stoter. 2008. SDI for public governance: Implications for evaluation research. In A multi-view framework to assess spatial data infrastructures, ed. J. Crompvoets, A. Rajabifard, B. Van Loenen, and T. Delgado Fernández, 51–68. Wageningen, The Netherlands: Space for Geo-Information (RGI).

Giff, G. 2008. A framework for designing performance indicators for spatial data infrastructure assessment. In *A multi-view framework to assess spatial data infrastructures,* ed. J. Crompvoets, A. Rajabifard, B. Van Loenen, and T. Delgado Fernández, 211–234. Wageningen, The Netherlands: Space for Geo-Information (RGI).

GOI-DST (Government of India, Department of Science and Technology) Task Force on NSDI. 2001. National spatial data infrastructure (NSDI) strategy and action plan. Discussion document ISRO-NNRMS-SP-75-2001. http://nsdiindia.gov.in/nsdi/nsdiportal/images/NSDIStrategyActionPlan.pdf (accessed January 15, 2010).

Harvey, F. 2006. Elasticity between the cadastre and land tenure: Balancing civil and political society interests in Poland. *Information Technology for Development* 12:291–310.

Harvey, F., and D. Tulloch. 2006. Local-government data sharing: Evaluating the foundations of spatial data infrastructures. *International Journal of Geographical Information Science* 20:743–768.

Heeks, R. 2002. Information systems and developing countries: Failure, success, and local improvisations. *The Information Society* 18:101–112.

Homburg, V., and Y. Georgiadou. 2009. A tale of two trajectories: How spatial data infrastructures travel in time and space. *The Information Society* 25:303–314.

Janssen, K. 2008. A legal approach to assessing spatial data infrastructures. In *A multi-view framework to assess spatial data infrastructures,* ed. J. Crompvoets, A. Rajabifard, B. Van Loenen, and T. Delgado Fernández, 255–272. Wageningen, The Netherlands: Space for Geo-Information (RGI).

Kitchin, R., C. Perkins, and M. Dodge. 2009. Thinking about maps. In *Rethinking maps: New frontiers in cartographic theory,* ed. M. Dodge, R. Kitchin, and C. Perkins, 1–25. London: Routledge.

Klein, H. K., and M. D. Myers. 1999. A set of principles for conducting and evaluating interpretive field studies in information systems. *Management Information Systems Quarterly* 23:67–94.

Lance, K. T. 2008. SDI evaluation and budgeting processes: Linkages and lessons. In *A multi-view framework to assess spatial data infrastructures,* ed. J. Crompvoets, A. Rajabifard, B. Van Loenen, and T. Delgado Fernández, 69–92. Wageningen, The Netherlands: Space for Geo-Information (RGI).

Lance, K. T., Y. Georgiadou, and A. Bregt. 2006. Understanding how and why practitioners evaluate SDI performance. *International Journal of Spatial Data Infrastructures Research* 1:65–104.

Madon, S. 2009. *e-Governance for development—A focus on rural India, technology, work and globalization.* Basingstoke, England: Palgrave Macmillan.

Masser, I., A. Rajabifard, and I. P. Williamson. 2007. Spatially enabling governments through SDI implementation. *International Journal of Geographical Information Science* 22:5–20.

Mol, A., and J. Law. 2002. Complexities: An introduction. In *Complexities: Social studies of knowledge practices,* ed. J. Law and A. Mol, 1–22. Durham, NC: Duke University Press.

Nedović-Budić, Z., J. Pinto, and N. Raj Budhathoki. 2008. SDI effectiveness from the user perspective. In *A multi-view framework to assess spatial data infrastructures,* ed. J. Crompvoets, A. Rajabifard, B. Van Loenen, and T. Delgado Fernández, 273–304. Wageningen, The Netherlands: Space for Geo-Information (RGI).

Pipek, V., and V. Wulf. 2009. Infrastructuring: Towards an integrated perspective on the design and use of information technology. *Journal of the Association for Information Systems* 10:447–473.

Prakash, A., and R. De'. 2007. Importance of development context in ICT4D projects: A study of computerization of land records in India. *Information Technology and People* 20:262–281.

Puri, S. K. 2006. Technological frames of stakeholders shaping the SDI implementation: A case study from India. *Information Technology for Development* 12:311–331.

Puri, S. K., S. Sahay, and Y. Georgiadou. 2007. A metaphor-based sociotechnical perspective on spatial data infrastructure implementations: Some lessons from India. In *Research and theory in advancing spatial data infrastructure concepts,* ed. H. Onsrud, 161–173. Redlands, CA: ESRI Press.

Rajabifard, A. 2008. A spatial data infrastructure for a spatially enabled government and society. In *A multi-view framework to assess spatial data infrastructures,* ed. J. Crompvoets, A. Rajabifard, B. Van Loenen, and T. Delgado Fernández, 11–22. Wageningen, The Netherlands: Space for Geo-Information (RGI).

Schuurman, N. 2005. Social perspectives on semantic interoperability: Constraints on geographical knowledge from a data perspective. *Cartographica* 40:47–61.

Schuurman, N., and A. Leszczynski. 2006. Ontology-based metadata. *Transactions in GIS* 10:709–726.

Silva, L. 2007. Institutionalization does not occur by decree: Institutional obstacles in implementing a land administration system in a developing country. *Information Technology for Development* 13:27–48.

Silverman, D. 1998. Qualitative research: Meanings or practices? *Information Systems Journal* 8:3–20.

Star, S. L., and K. Ruhleder. 1994. Steps towards an ecology of infrastructure: Complex problems in design and access for large-scale collaborative systems. *CSCW 94 ACM Conference on Computer Supported Cooperative Work,* 253–264. New York: ACM Press.

State Council for Science and Technology. 2008. Development of Web based geoportal and data clearing house progress report. http://nrdms.gov.in/minitues-reports/second-meeting/kscst-on-pr.pdf (accessed January 25, 2010).

Steudler, D., A. Rajabifard, and I. Williamson. 2008. Evaluation and performance indicators from an organizational perspective. In *A multi-view framework to assess spatial data infrastructures,* ed. J. Crompvoets, A. Rajabifard, B. Van Loenen, and T. Delgado Fernández, 193–210. Wageningen, The Netherlands: Space for Geo-Information (RGI).

Vandenbroucke, D., K. Janssen, and J. Van Orshoven. 2008. INSPIRE state of play: Generic approach to assess the status of NSDIs. In *A multi-view framework to assess spatial data infrastructures,* ed. J. Crompvoets, A. Rajabifard, B. Van Loenen, and T. Delgado Fernández, 145–172. Wageningen, The Netherlands: Space for Geo-Information (RGI).

Van Loenen, B., J. Besemer, and J. Zevenbergen. 2009. Spatial data infrastructure convergence. In *SDI convergence—Research, emerging trends, and critical assessment,* ed. B. van Loenen, J. W. J. Besemer, and J. A. Zevenbergen, 1–7. Delft: NCG.

Walsham, G. 2006. Doing interpretive research. *European Journal of Information Systems* 15:320–330.

Walsham, G., D. Robey, and S. Sahay. 2007. Foreword: Special issue on information systems in developing countries. *Management Information Systems Quarterly* 31:317–326.

9

Considerations from the Development of a Local Spatial Data Infrastructure

Clodoveu A. Davis, Jr. and Frederico Fonseca

CONTENTS

ABSTRACT The development of spatial data infrastructures (SDI) is often the victim of an excessive focus on data and standards. In this paper we use the hermeneutics of Gadamer and Habermas to understand the problem of how SDIs can succeed. We use Gadamer's concept of phronesis to show how being an application-driven project is a key for success, and Habermas' ideas to show the importance of emancipatory knowledge in the implementation of SDIs. A case study on a GIS project in Belo Horizonte, Brazil, is presented. The project has been evolving for fifteen years, although it started with a focus on data and standards, and generated a strong and active spatial data infrastructure for the city. The reasons for success were manifold, but we highlight the application-driven nature of the project, along with the combination of multiple disciplines and multiple levels of expertise in its design and implementation team.

KEYWORDS: Spatial data infrastructures, geographic information systems, cooperation.

9.1 Introduction

In the call for papers for this special issue, Georgiadou et al. (2004) warn us that both theory and practice of current research into SDIs focus "strongly on issues related to spatial data and their accessibility (for example, issues of standards and interoperability)" and contribute "significantly to the unrealized benefits of SDI." Although we agree with them in that a strong focus on data is a frequent cause of failure as can be seen in the GIS literature, we present in this paper a project that succeeded in a transitional economy (Brazil) despite having started with an enormous investment in data and with its first project being directed to data exchange. Why did such a project succeed? What were the causes that avoided failure and led to a strong and thriving spatial data infrastructure? These are the questions that this paper will address based on two assumptions. First, we understand that knowledge is decisive to the development of transitional economies. Second, we discuss a hermeneutic view that knowledge, by itself, without an application, is meaningless. The knowledge that is needed by transitional economies is applied knowledge. Notice, however, that this problem also occurs in developed economies, where sustainability of IT projects is not a trivial issue. This paper will show that many solutions that have been adopted in Brazil could also be applied in developed countries.

There is a consensus that knowledge is decisive for development. Economies are created and sustained on the foundation of information, learning, and adaptation, and not only with the accumulation of physical capital. Nevertheless, creating, acquiring, and using scientific and technological knowledge in developing countries is a Sisyphean task (Sagasti, 2004). A key component in the implementation of information and communication technologies (ICTs) in transitional economies is what Braa et al. (2004) call sustainability. It is "the challenge to make an information system work, in practice, over time, in a local setting. This involves shaping and adapting the systems to a given context, cultivating local learning processes, and institutionalizing routines of use that persist over time (as well as when the researchers leave and external funding is over)" (p. 338). One of the requirements for the sustainability of technological projects in developing countries is that indigenous knowledge and techniques are respected, maintained and included in the process of implementation of new technologies. The new technologies have to be seized upon and appropriated if they are to serve the goals of social and human development in these countries. Otherwise, the knowledge paradigm will only increase the already alarming levels of exclusion and inequality (Reed, 2000).

We argue that Belo Horizonte's GIS project (GIS-BH), our case study, is such a case, in which knowledge was appropriated and applied to local conditions. In order to understand why it succeeded we will use Gadamer's hermeneutic approach to knowledge, presented in *Truth and Method* (Gadamer, 1975), and Habermas' three types of knowledge, as laid out in *Knowledge and Human*

Interests (Habermas, 1971). In his work, Gadamer studies Aristotle and the concept of *techne,* the knowledge "of a craftsman that is able to make some specific thing" (p. 281). But a broader knowledge, one that can be applied to social situations, has to be a knowledge of a dynamic kind, one that determines and guides action. Consequently it "must include the application of knowledge to the particular task" (p. 281). We will argue here that, in the project we study, applications were the driver of everything. Although the project had an early focus on data and standards, the fact that it was *application-driven* made it successful. We define applications here in the hermeneutic sense, which is broader than the software application sense, although both are related. Gadamer considers "application to be as integral a part of the hermeneutical act as are understanding and interpretation" (p. 275). We apply to the understanding and assimilation of technology what he is saying about the understanding of texts. "This includes the fact that the text (the technology, our parentheses), whether law or gospel, if it is to be understood properly, i.e., according to the claim it makes, must be understood at every moment, in every particular situation, in a new and different way. Understanding here is always application" (p. 275).

We will discuss the development of a local SDI from the perspective of the GIS-BH project, from the city of Belo Horizonte, Brazil. It is a nationally recognized initiative, known for its pioneering nature and innovative proposals, with an emphasis on social applications. GIS-BH also stands out for having established, early in the project, a cooperation agreement with numerous partners for costless data exchange and distribution. Such a broad cooperation increased the quantity and the variety of applications, thus exerting a stabilizing effect and pushing the GIS towards becoming a spatial data infrastructure (SDI) on urban data. The bottom-up, applications-driven approach that led the project to achieve the present stage of development is the focus of this paper.

This paper is organized as follows. Section 9.2 presents a brief theoretical discussion on the process of knowledge acquisition and sustainability, which is central to the arguments presented here. Section 9.3 presents the development of GIS-BH, pointing out elements and key events that can help in understanding its success. Section 9.4 presents the link between our theoretical concepts and the real project. Section 9.5 presents a set of recommendations for similar projects. Finally, Section 9.6 presents our conclusions and indicates directions for future work on this subject.

9.2 Moving from Techne to Phronesis

Central to the hermeneutic approach proposed by Gadamer, and very important for our analysis of the acquisition and sustainability of knowledge in a transitional economy, is the fact that a given tradition or practice must be understood again and again, depending on the context and situation. A common situation in the attempts to establish SDIs in transitional economies is the use of packages funded by international agencies. This kind of solution makes an intensive use of foreign expertise, which often lacks a deeper knowledge of local conditions.

Understanding, Gadamer says, is "a special case of applying something universal to a particular situation" (p. 278). In order to understand this problem, Gadamer uses the concept of *phronesis* from Aristotle. Practical knowledge, or phronesis, is a special kind of knowledge, as opposed to *techne* and *episteme*. Phronesis is knowledge directed towards a concrete situation, while episteme is "scientific knowledge, knowledge of what is universal, of what exists invariably" (Bernstein, 1983). Techne corresponds to technical know-how and can be learned and forgotten, while ethical reasons cannot. Gadamer's intention was to show that knowledge cannot be understood by itself. Instead, it must not be "detached from a being that is becoming" but it is "determined by it and determinative of it" (p. 312).

Also related to Gadamer's position is the idea of knowledge as a fundamental component in development. A World Bank report (The World Bank, 1999) suggests two views of knowledge: first, knowledge about technology, technical knowledge or know-how. This is the software engineering aspect of SDI. Second is knowledge about attributes. This is knowledge about the needs and local conditions. The lack of the latter generates "information problems" (The World Bank, 1999). This perspective of knowledge refers to the results of the use of technology. These results should include geographic information systems that address information issues of public health, transportation, security, and other matters of interest of developing countries.

Habermas (1971) has a more precise elaboration of knowledge and of how it can be used as an emancipatory instrument. For him, knowledge can be seen in three different categories. First, there is "information that expands our power of technical control" (p. 313). Second, there are "interpretations that make possible the orientation of action within common traditions" (p. 313). And finally, and more importantly, there are "analyses that free consciousness from its dependence on hypostatized powers" (p. 313). Habermas (1971) names the three categories of possible knowledge as *technical, practical,* and *emancipatory*. Emancipatory knowledge is achieved by the combination of the other two types of knowledge. We argue that emancipatory knowledge will be acquired by developing countries through an adequate use of the technology. Our expectation about the potential use of SDIs in transitional economies can also be seen as an extension of Habermas' position about the possible kinds of knowledge. We argue that in order to be successful in transitional economies, SDIs have to play the role of emancipatory knowledge.

This is precisely the case in GIS-BH, in which the need to stabilize the project, both politically and financially, along with the need to ensure continuity and sustainability, caused the project to be developed in a bottom-up and application-driven fashion. As the detailed account presented in the next section shows, even though the project's initial goals were ambitious, their realization took place over a long time—over fifteen years—through a succession of cautious, localized, and often negotiated actions that created and maintained a favorable environment for the development and expansion of the GIS and its use.

9.3 Belo Horizonte: From Mapping to GIS to SDI

Belo Horizonte is the fourth largest Brazilian city, with a population of more than 2.2 million people, spread over 335 square kilometers, and is the center of a metropolitan area that houses over 3.5 million people. Belo Horizonte's GIS development effort started in 1989 and has been continuously evolving ever since. Throughout its history, GIS-BH has received awards, has been widely presented in the form of invited lectures at many events nationwide, and has generated a body of over 200 publications, including dissertations and theses, academic papers with technological and methodological innovations, conference papers about applications development, and technology dissemination articles. The project's priorities lie in social applications, including education, health, transportation, traffic and environmental control, among many others. The authors of this paper have both been members of the GIS-BH team for many years and currently cooperate with their colleagues that are still working in the project through academic, technical, and personal links.

GIS-BH was created as part of the administration's response to the numerous new challenges presented by the 1988 Brazilian Constitution. The new constitution emphasized a shift of responsibilities from the federal and state levels to the local level. Various public services, such as health, basic education, water and sewage, energy, transportation, and traffic, were transferred from other levels of government to the municipalities, along with regulations and standards that placed additional pressure on them to be more responsive towards the demands of the citizens. Municipal governments were also expected to be much more accountable and transparent, providing detailed information to the public in order to allow the effective exercise of the democratic right of participation.

In the early nineties, many GIS projects were almost exclusively driven towards revenue increases, following a sales pitch presented by many vendors and consultants. According to these "specialists," deploying a GIS would cause enough revenue increases to pay for the entire technological investment in a very short time. In fact, methodological shortcomings and lack of consistent updating caused cadastral bases for taxing purposes to be quite outdated and flawed; thus data checking or updating efforts was likely to have caused revenue increases anyway—regardless of the use of GIS. For that reason, most urban GIS projects in Brazil were led by tax collection departments, rather than by urban planning departments, or even by IT departments.

In Belo Horizonte, GIS was developed at the municipal IT company (Prodabel), which was also responsible for the city's cartography. This unusual scope of activities enabled Prodabel, early in the project, to form a team of specialists in several IT areas (databases, information systems, computer graphics) as well as in fields such as urban cadastre, cartography, surveying and others.

Concern on multiple uses of the data was present early on the project. Ensuring the level of investments and political support required to push the project forward, at a time when this technology was largely unknown, required project managers to propose applications in many different areas. Project managers were also able to convince decision makers throughout the administration that a solid base map was required in order to provide adequate support for the thematic applications.

With this, Belo Horizonte's GIS faced, early on, three important challenges: (1) building a general-purpose database, (2) developing a wide range of applications, mostly in social fields, and (3) keeping this database up-to-date, as required by the applications (Davis Jr. & Zuppo, 1995).

The first challenge is directly related to research in topics such as data transfer standards, evolving towards interoperability, and then on to semantics and ontologies. This caused part of Prodabel's GIS team to evolve into a research team, continuously seeking innovative approaches and solutions to all these themes, often in cooperation with universities and research centers. (See Davis Jr., 1995, 2002; Fonseca & Davis Jr., 1999; Fonseca, Egenhofer, Davis Jr., & Borges, 2000; Fonseca, Egenhofer, Davis Jr., & Câmara, 2002, for a sequence of research initiatives conducted along that path.)

The second challenge regards arguably the most important aspect of GIS as a technological tool, which is its interdisciplinary nature. From this, the involvement of specialists from each application area was required, thus forcing the establishment of strong connections between Prodabel's original GIS team and thematic specialists in each of the city's departments, particularly in health, education, sanitation, transportation, planning and licensing. In each of these areas, GIS became a tool geared towards technical activities, used directly by technicians, with IT and base map support by Prodabel.

The third challenge required an approach being neither academic (as the first) nor integrational (as the second). Updating such a varied database (currently comprising over 6 million objects, distributed through over 300 object classes) required a strong coordination of efforts and cooperation with external agencies (utility companies, state government departments, federal institutions, universities, and others). In turn, this drive towards cooperation provoked an interest in data sharing among the municipal administration and these external actors, which expanded even more the range of GIS data and applications, thus forming a virtuous circle, leading to increased data quality, interoperability, and scale gains.

In this section, we describe the development of Belo Horizonte's GIS since 1989, showing in detail how these challenges (and others) were met, leading to what can be construed as the foundations of a true SDI for the city. We divide this period into three phases. The first phase corresponds to the *initial capacity building*, including initial data set creation, hardware and software acquisition, personnel training, and initial applications development. In the second phase, the GIS acquires more maturity, shifting the focus to *sustainability* through the deployment of a wide range of applications and the assurance of data quality through maintenance routines. The third phase corresponds to the *maturity* of the GIS, a phase in which the accumulation of knowledge and experience with urban GIS leads to a more secure definition of goals and to a vision of the role this technology can play in the future of local government. In this phase, a new technological architecture (Davis Jr. & Oliveira, 2002) starts to replace hardware and software from the early 1990s, data sets are renewed, and an already wide cooperation agreement gains further momentum. Each of the phases will be described in detail through the next subsections.

9.3.1 Initial Capacity Building (1989–1992)

This phase began as soon as the decision was taken to create a new base map, in electronic form, for the city of Belo Horizonte. This base map was created from a new aerial survey, followed by extensive and detailed stereoplotting. At that time, this process was both costly and slow: It took more than two years to be complete and cost over US$1.00 per inhabitant.

Figure 9.1 shows, schematically, how this first phase evolved, considering three main aspects: development of the initial geographic database (GDB), establishment of cooperation, and technology acquisition.

While the stereoplotted data were being developed, Prodabel started a series of discussions with institutions and companies that were deemed as possible partners in the use and updating of the base map. This included the utility companies (power, water and sewage, telecom) plus several municipal, state, and federal organizations. In these discussions, the possibility of sharing the data resulting from the rather large investment in database creation acted as a catalyst to push cooperation forward. Even before a formal cooperation agreement was signed, some of the interested parties were already working informally with Prodabel, in order to develop a data set that was deemed as strategically important to all participants, and that would not result from the stereoplotting process: an address database. First, cooperation involved generating a compatibility table for street codes. Later, this evolved into the development of our first common address data set, with over 300,000 individual addresses, georeferenced as points (Davis Jr., 1993). The creation of the address database was further accelerated by the creation of an image data set by the scanning of the existing analogous cadastral plans. This element also helped us to accelerate the transition from the previous cadastral routines to the new system.

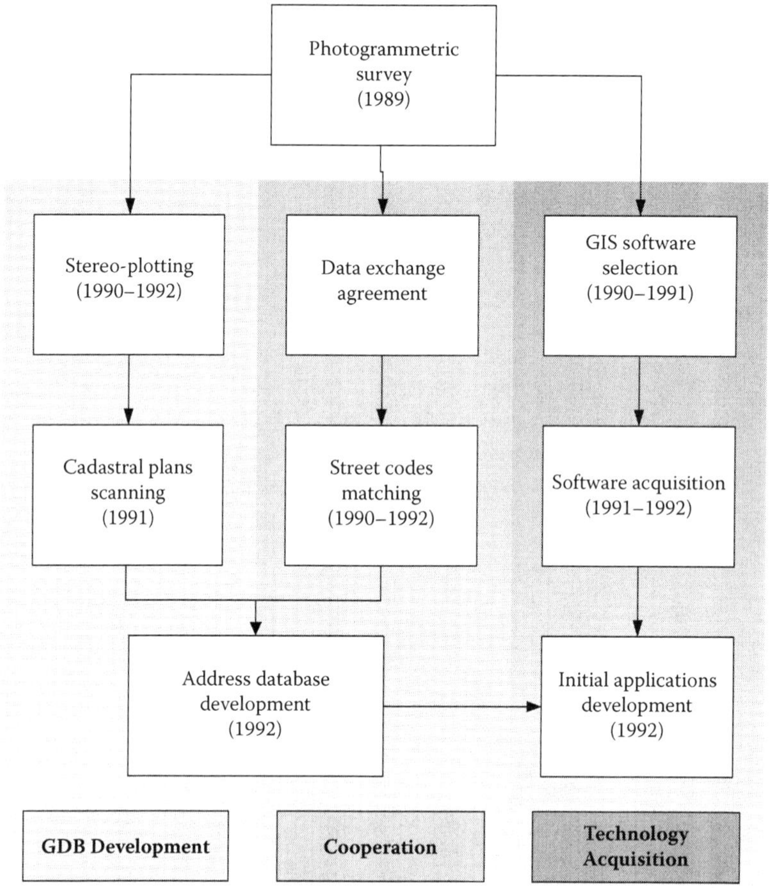

FIGURE 9.1
Initial capacity building.

Also while the stereoplotted data were being generated, Prodabel's GIS group took on the creation of a request for proposals (RFP) for the acquisition of GIS software. In this RFP, technical features of existing products were compared to the project's expectations. Due to the lack of knowledge and to the lack of a fair ground for feature-by-feature comparison, the RFP was finally published containing a very broad spectrum of good intentions, most of them application-based, and asked vendors for comments on how to accomplish those ideas using their products. As a result, Prodabel purchased a French GIS software (APIC), which was based on an object-oriented database, had a powerful query language, and included a number of innovative features (Davis Jr. & Borges, 1994). Prodabel has been the sole user of that software in Brazil ever since. The need to communicate with other GIS users eventually led Prodabel's team to create a drive towards a more complete understanding of the fundamentals of GIS, therefore going much further than it would have been possible by simply learning the operational aspects of the software.

The installation of hardware, the training in APIC and the delivery of the initial data set from stereoplotting took place almost at the same time—late 1991 and early 1992—just before a political change in Belo Horizonte's municipal administration, mandated by the October 1992 elections. This political factor was very important, since the large investment in data, hardware, and software, along with the wide discussions inside and outside the municipal administration, led to high expectations. Also, since the mayor was a former IT specialist, and thus able to fully understand the complexity of this process, the project had been quite well insulated from political turbulence—up to that point. Because of that, some initial applications were developed mostly to serve as "demos," to display the possibilities of this new technology. These applications were widely showcased, at every possible opportunity (conferences, expositions, political events, and so on), and even received a national award at a major environmental event.

At that time, however, even though all of the project's initiatives were working properly, some obstacles that could affect the progress of the project were becoming quite clear. First, the proprietary GIS solution purchased by Prodabel had a number of advantages on the technological side, but was somewhat difficult to integrate with other information systems. Also, since Prodabel was the only user, there was no source from which the company could hire development and training services. The solution envisioned at that time was to increase the investment in Prodabel's own personnel, and to identify as soon as possible a desktop GIS alternative to APIC, in order to increase the *capillarity* of the GIS with reasonable costs. We opted to use the term "capillarity" to describe how GIS was spread out through the administration in our project, since it conveys a sense of "penetration, in virtue of the porosity," as defined by the Oxford English Dictionary, as opposed to "dissemination," by which it could be incorrectly understood that the same kinds of tools are supposed to be used throughout the administration. Capillarity means reaching out to the most remote (and often information-thirsty) post in the local government, either through small GI applications or by just providing access to data.

Another problem was the risk for investment discontinuity, mostly because of the upcoming elections. If investments failed at that phase of the project, serious problems could arise from the rapid deterioration of data quality and difficulties to keep up with technological evolution. To counter this problem, a strategy to establish a firm foothold for the GIS in the municipal administration, through a wide diversity of applications, was established. Furthermore, specialists in each application's field were summoned to participate in the development efforts. At that stage, it was evident that a clear strategy for sustainability was required, thus inaugurating a new phase in the project.

9.3.2 Sustainability Phase (1992–1995)

The first response for the sustainability challenge took place intuitively, in the form of the cooperation agreement, which was under development while the first digital data set was being prepared. Even though an existing agreement with numerous external entities would help, sustaining the project evidently required a continuing expansion of the user base, and in turn that would only be achieved through more applications.

Figure 9.2 shows how this phase occurred, considering again three different aspects, as in Figure 9.1. First, the GDB construction went on, adding to the database a number of object classes that were important in a multi-applications environment, and corresponded to data that could not be collected by aerial photogrammetry. These included the most important elements of the urban infrastructure and services, street centerlines, and spatial reference units of all kinds (neighborhoods, health care districts, census sectors, planning sectors, and so on). The second aspect evolved from an emphasis on cooperation to some actions developed with the purpose of promoting the system's consolidation, through the widest possible dissemination. Prodabel's GIS team started

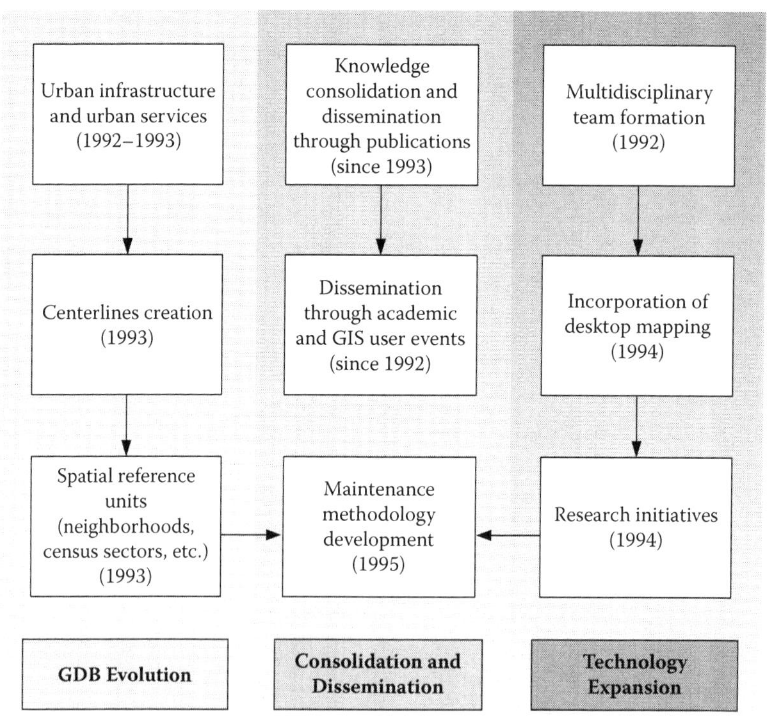

FIGURE 9.2
Sustainability phase.

to participate in many GIS conferences, nationally and internationally, and to publish articles on GIS concepts, development strategies, technology, maintenance, and other aspects in all sorts of publications. Regarding technology acquisition, in this phase Prodabel's multidisciplinary team became fully operational and started to expand its reach through new technologies (such as desktop mapping, remote sensing, digital image processing, and others). Research initiatives began to take place, motivated by the early publications and participation in academic events and fueled by the team's technical quality.

A fundamental landmark of this phase is the development of a maintenance methodology for the digital data (Davis Jr. & Zuppo, 1995). This methodology was developed as an adaptation of time-tested manual, paper-cartography-based routines, to incorporate GIS as a tool (Silva & Ottoni, 1995). Since Prodabel's cadastral team had a large experience in field work, the maintenance methodology was created with most of the correct concerns in mind. The most important decision here was on prioritizing the maintenance of data that would potentially be shared among numerous applications. Data that were merely "cartographic," meaning object classes that were vectorized from aerial photos solely to fulfill cartographic standards, were not considered until some application determined its use and the adequate method for its maintenance. The development of this methodology serves as a classical example of the fusion of practical and technical knowledge, leading successfully to emancipatory knowledge.

In this phase, some data-related initiatives stand out as early attempts to reach a higher level of integration between Prodabel and its partners. Members of the cooperation agreement conceived and agreed on a standard for data classification, which served as a base for a broad information catalog. This experience led, later on, to a state-wide geoinformation assets catalog. In this classification, data were regarded as being *essential*, if the information is required for the operation of a given organization, or *complementary*, if the information is useful for the organization, but is not mandatory for its activities. Parallel to that, data were classified as *specific*, if its generation and maintenance are part of the organization's institutional responsibility, or as *common*, if the information is usually obtained from external sources. There was also a classification based on levels of confidentiality and privacy (Davis Jr., 2002).

Regarding standards, in this phase an attempt was made to create a data interchange standard for the cooperation agreement, a data storage format called IIG. This standard was neutral in respect to existing GIS software, so that it would only be necessary for each partner to develop converters to/from their GIS from/to IIG, instead of relying on specific converters, developed for each pair of existing GISs, or on transferring data encoded in CAD format (Davis Jr., 1995, 2002).

Agreements on data sharing were also reached in this phase. First, an agreement on street codes and addresses made it feasible to create a street code conversion table involving all the coding systems in use (Prodabel, power company, water and sewage company, telecommunications company, police, and postal codes). As a result, many of the partners revised their conventional information systems to reflect and to facilitate this integration. An agreement on the boundaries of some commonly used spatial information units was also developed in this phase.

To summarize, the sustainability phase was characterized by numerous efforts to stabilize, standardize, consolidate, and evolve with the GIS towards its original goals. This was achieved by deepening the emphasis on applications, by increasing the GIS team's technical capacity, and by extending the reach of the GIS data to as many partners and users as possible. Making good use of the investment in the GIS has driven these efforts, thus generating a successful strategy of turning something that was seen as *expensive* into something perceived as inherently *valuable*, inside and outside the municipal administration.

9.3.3 Maturity Phase (1995–Today)

The next phase of the project is characterized by extensive usage of GIS resources and by a drive towards a new technological architecture (Davis Jr. & Oliveira, 2002), through which the distribution of data and the integration with partners can be achieved with greater ease and flexibility.

Figure 9.3 shows a diagram of the main actions in this period. We can now consider database actions to be geared towards integration, since they mostly include general-interest items, such as a new high-resolution imagery set. We can also start mentioning definite sustainability efforts, which in this phase include greater capillarity through the use of desktop mapping and, hopefully in the near future, Web GIS. A temporary solution to rapid and non-bureaucratic dissemination of data was the creation of a FTP server containing most of the shared data, with free access to municipal organizations and partners. This server was initially at Prodabel, but it has recently moved to a machine installed at Minas Gerais Federal University.

Regarding the technological aspects, in this phase we can actually say that Prodabel's research and development efforts started producing results, materialized as methodologies, in-depth studies, and experiments with issues that are clearly in the project's future. A geographic data modeling technique was proposed and is currently used in many organizations throughout the country (Borges, Davis Jr., & Laender, 2001). A new architecture for the GIS was studied and is currently under implementation. This new architecture incorporates elements that allow it to be interoperable and distributed, and with a strong support for digital imagery (Davis Jr. & Oliveira, 2002). Specialists at Prodabel are currently researching advanced subjects, such as process and action modeling through ontologies, visualization in large spatial and spatio-temporal databases, and service-oriented architectures.

Another maturity aspect is the scope and range of applications. GIS-BH started the millennium with several working applications, not mentioning punctual efforts and support to numerous city projects. Most of the

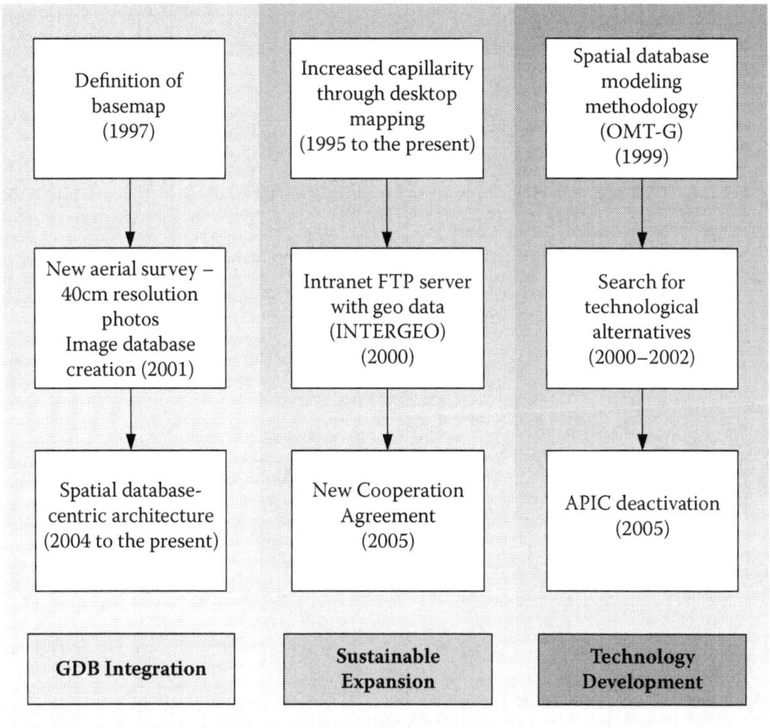

FIGURE 9.3
Maturity phase.

applications cover social fields and include, but are not limited to the following:

- **Education:** distribution of students among grammar schools based on spatial criteria. A pre-registration, in which the home address is recorded for each student, precedes the actual enrollment in a school. Based on the address, a school for enrollment is assigned to each student (Fonseca & Zuppo, 1994; Pinto, 1999).
- **Health care:** monitoring of epidemics and diseases propagated by vectors; monitoring of health-related social indicators (Pessanha & Carvalho, 1999; Sousa & Bretas, 1994).
- **Transportation and traffic:** creation and maintenance of the vehicle circulation network, including all circulation constraints; recording the location and contents of every traffic sign in the city; bus itineraries and stops; traffic accident mapping (Meinberg, 2003; Zuppo, Davis Jr., & Meirelles, 1996).
- **Sanitation:** recording of all garbage collection and street sweeping routes; database of all public trash cans.

- **Land use and zoning:** a database containing the city's land use zones and associated parameters.
- **Parceling:** reparceling and validation of illegally parceled areas; tools to compare actual parceling with proposed/approved parceling (Peixoto, 2000).
- **Urban planning:** support for the development of Belo Horizonte's Master Development Plan and the creation of socio-economic indicators (S. M. Oliveira, Sousa, Davis Jr., & Amaral, 1996).
- **Public safety:** crime mapping and analysis (by Minas Gerais State military police, PMMG).
- **Water and sewage:** complete networks (by Minas Gerais State water and sewage company, COPASA-MG).
- **Drainage:** complete network, with engineering and maintenance details.
- **Power distribution:** complete network (by Minas Gerais energy company, CEMIG).
- **Public lighting:** every lighting fixture's position recorded, along with lamp type.
- **Favelas (slums):** mapping, support in the creation of development plans.
- **Economic activities:** mapping, planning, concentration analysis (M. P. G. Oliveira, 1997).

With such a wide variety of applications and a significant number of partners, inside and outside the municipal administration, there was a clear need for a better arrangement as to the maintenance tasks. Prodabel remained in charge of coordinating the updating of every information class that is the potential object of interest from multiple users. With this, a decision was reached as to the contents of a "base map," or rather, a "basic information set." Basically, this basic information set includes address data, urban cadastre, most spatial reference units, digital cartographic data, demographics (S. M. Oliveira, Sousa, & Amaral, 1995), and imagery. Other municipal organizations or partners are in charge of updating information that is specific to their institutional responsibility.

With this, there are actually multiple updating efforts taking place in Belo Horizonte at the same time. For instance, a municipal company in charge of slum support and urbanization (URBEL) is currently acquiring a new imagery set with a 10 cm resolution. CEMIG, the state-owned power company, is revising a database on public lighting, after a process to replace mercury vapor with the more economical sodium light bulbs. Both databases are to be shared with other partners at the end of their development.

However, the maturity of GIS-BH is better demonstrated by looking at the cooperative efforts to maintain an important part of the basic information set: the address database. This cooperation started even before the GIS, with the creation of the shared table of street codes we mentioned before. Currently, a group of professionals, representatives of every one of the 27 organizations that signed the cooperation agreement, meet every two weeks to discuss possible actions, to exchange knowledge on technical aspects, and to coordinate joint efforts. One of these efforts, recently concluded, involved several municipal organizations, along with the Brazilian census bureau (IBGE), the utility companies, the Brazilian postal services and a branch of the judiciary in order to revise and rebuild the addressing system in some favelas. This was performed with a very low investment, and extensive sharing of resources, including human resources: Data collection was performed by health agents and judiciary police officials; training and methodological orientation was provided by IBGE; digitizing was performed by Prodabel; and other partners contributed with items such as bus tickets, vehicles, plotting and other items.

Currently, the cooperation efforts are backed by a new version of the original agreement, which was signed in 1993. This new version essentially preserves the terms and definitions of its predecessor, and incorporates a new and easier process for the admittance of new partners. This agreement has been signed by 27 different organizations, at the municipal, state and federal governmental levels, plus private organizations that work on socially relevant fields and/or maintain shareable information on the city, which includes all privately owned utility companies.

This kind of data, expertise, effort, and investment sharing initiatives, along with the wide availability of general-purpose data, in standard formats, available on-line, and housed at a "neutral" server, motivates us to begin thinking in terms of a SDI for Belo Horizonte. Actually, the cooperation (and the interdependence that results from it) among multiple partners, to support multiple applications, with multiple clients, correspond to what is expected of a real SDI. Technological barriers were once great; at the time every partner used a GIS from a different vendor. With new technological tools, such as spatial databases and interoperable Web-based GIS, these barriers started to diminish in importance, and future pro-SDI elements (we can mention research topics such as ontology-driven GIS, geospatial semantics, service-oriented architectures, OpenGIS standards, geospatial Web services, and many more) will reduce those barriers even further.

9.4 Analysis: The Link between Theory and Practice

In this section, we show how Habermas' three categories of possible knowledge, technical, practical, and emancipatory, are linked to the GIS-BH project. Then we discuss how application, in the hermeneutical sense, was fundamental in the project. We show where Gadamer's concepts of techne, episteme, and phronesis are to be found in our case study.

9.4.1 Habermas' Emancipatory Knowledge in GIS-BH

Habermas (1971) came up with his three types of knowledge when he was studying a way of preserving the theoretical aspects of science while at the same time keeping its roots on practical aspects. He says, "Knowledge of the apparently objective world of facts has its transcendental basis in the prescientific world. The possible objects of scientific analysis are constituted a priori in the self-evidence of our primary life world" (p. 304). "Once the theoretical attitude has been adopted, it is capable in turn of being mediated with the practical attitude" (p. 305).

We start our analysis with practical knowledge. We consider that the knowledge about the data can be compared with practical knowledge. As we showed, the GIS project of Belo Horizonte started within the city department that was in charge of the city's cartography and urban cadastre. There was a long tradition of development and maintenance of cadastral information, even though most of it was maintained in non-digital form before the GIS deployment. Know-how on how to handle the data itself was already present. There was knowledge on how to do things in terms of geographical data.

The second type of knowledge analyzed by Habermas is technical knowledge. This kind of knowledge was present in two components of the GIS-BH project. The first one is the scientific education of the team in charge of the GIS project. All of them had a college education with degrees in engineering, computer science and business. Second, the project invested in science since its beginning, leading, years later, to a range of academic activities that include invited lectures at national and international meetings, publications, and even a professional masters program.

Our third and final component is emancipatory knowledge. The IT company, as well as its cartography and urban cadastre department, had a tradition of service prior to the start of the GIS project. Both had, as part of their basic mission, to fulfill and support the information needs of Belo Horizonte's local government and its citizens. Thus, we can relate our third knowledge component to the applications developed by the IT company and to the services provided by the cartography department, which later became a GIS department. It was the coexistence of practical and technical knowledge in the same place that generated the necessary balance to the creation of emancipatory knowledge.

9.4.2 An Application-Driven Project

Gadamer considers that understanding and interpretation, together with application, comprise one unified process. There can be no appropriation and understanding of technology without all three elements being present. We think that what leads many projects to fail and get stuck in the data and standards trap that Georgiadou et al. (2004) mention is the lack of applications. The development of a large and expensive geographic database without a specific focus on information needs most often drives down to a dead-end. In GIS-BH, this drive towards applications was present since the beginning.

Application is the key for knowledge that is engaged and leads to development. The path to this kind of knowledge has to include both practical knowledge and theoretical knowledge. In our case study, techne corresponds to the knowledge held by the technicians that dealt with cartography and urban cadastre, including the creation of digital maps, data maintenance, and the release of a final product to the public. Episteme is represented by the team that was in charge of studying and deploying the GIS, formed by specialists in various fields. In order to achieve phronesis, a surrounding background had to be present. This role was played by the emphasis on applications, with the obvious help of the fortunate combination of cartography, urban cadastre, and IT in the same corporate environment, which proved to be a decisive factor. Applications provided a common background for these teams, guiding the data acquisition and treatment efforts as much as guiding knowledge acquisition and technological development.

The application-driven nature of the project can be seen in three examples. First, Prodabel was a company that had as a mission to develop applications for the local government of Belo Horizonte. Therefore, application was in its nature. All the technology developed and controlled by the company had only one objective, which was to provide services to the city.

Second, the cartography and urban cadastre department had also a mission of service. It was the main source of cartographic data for the city. It had a longstanding tradition of providing maps that include applications ranging from solving legal disputes to address matching and the record of the distribution of physical resources.

Finally, the first large application of the project after that creation of the database, namely the compatibilization of street codes, required the establishment of a standard for data exchange. Here the difference was that the standard was not a thing in itself, and had a very direct purpose. The main objective was the exchange of home and business addresses. The focus on real data that was used to complement each other's databases did not let the standard become ineffective. On the contrary, all the participants were engaged and actively exchanged data and helped update each other's databases, generating a good starting point for other cooperation initiatives.

9.5 Lessons Learned: Recommendations for Similar Projects

The GIS-BH project has become successful due to a combination of factors. Some of these factors are related to local characteristics and political context. In particular, the need to update the municipal cartography, simultaneously with the political interest in developing the city's information systems, provided a good opportunity for the creation of a project with broad objectives. Funding was also facilitated in this politically favorable context, and the need for political justification for expenditures in IT and database formation helped to guide the project towards application-specific goals, thus making it clear that IT is being regarded as means to an end. We understand that, while these conditions may not be immediately available in every context, there must be a serious concern with (1) funding sources, (2) political visibility, and (3) application-based justification for the investment.

Some factors of success, however, are general enough to serve as a basis for the proposition of a set of recommendations and assessment criteria for similar projects, both ongoing and initiatives that have to start from scratch. We divided these recommendations into three categories: *applications, cooperation,* and *people.*

- **Applications:** When determining the information needs of applications, focus on data that can potentially be shared among other applications. The maintenance of such data is likely to be critical. The data requirements of the applications must drive the maintenance efforts. Maintenance methodology must be developed from previous (non-digital) practices, using local knowledge, experience, and expertise. The intended users for the applications must also participate in the process, as specialists in the field related to the application (see below).
- **Cooperation:** From the selection of applications and the determination of their information needs, identify potential partners for data sharing and cooperation in the development of data sets. Such partners will conceive their own applications, and the inter-organizational demand for some information classes will be important for the stability of the entire information system. Partners should not be forced to use the same technological tools or products; this decision should take into consideration the specific needs of the applications envisioned by the partner. Data sharing should not be format- or product-specific. If at all possible, sharing of existing data should be at no cost. Cost sharing for further data collection needs to take into consideration maintenance costs and effort, which may also be shared; this means that an organization that commits to provide maintenance to a class of data can be relieved of their share of the cost of initial data collection. It also means that cost-sharing agreements should be made considering aspects of budgeting, investment capacity, political calendar and others that are specific to each participating organization. In a formal multilateral cooperation agreement, rules should be flexible enough to allow the inclusion of new partners.
- **People:** Team formation and preparation must take into consideration that most GIS- and SDI-related efforts are multidisciplinary.

> Therefore, teams must include specialists from different fields, and the flexibility to admit new members depending on the requirements of each task. Leadership within the team must also be flexible, in order to give the specialist in the field that is closer to the problem at hand more leeway in the search for its solution. Team members must not only focus on learning how to operate the tools, but also need to understand the scientific and technological foundations of GIS and SDI. It is also important to systematically record and disseminate the results achieved in each step of the way, both internally and externally, to reinforce political support.

We observe that projects that have definite application concerns and clear objectives as to their societal contribution have less difficulty in obtaining funding, gaining political support, and achieving the kind of visibility that ensures continuation. We also observe that projects that materialize intra-departmental or inter-organizational cooperation have a lower risk of discontinuity for political or financial reasons. Therefore, projects that are ongoing or that are about to be started should consider reviewing their objectives and intended relationships with other organizations with a focus on sustainability, considering the points we presented here.

9.6 Conclusions and Future Work

In this paper we described a GIS project, GIS-BH, which, even though it has started with a focus that often leads to failure (data and standards), managed to survive and end up generating a strong and active spatial data infrastructure for the city of Belo Horizonte, Brazil. The reasons for success were manifold, but we highlighted here the application-driven nature of the project. Also, the combination of multiple disciplines and multiple levels of expertise (ranging from practical to theoretical) was fundamental for the generation of applications and agreements that answered the information needs of the local community. We used Gadamer's and Habermas' insights into understanding and knowledge to explain some of the reasons that made the GIS-BH project become a foundation for Belo Horizonte's SDI.

While participating in the development and expansion of GIS-BH, we have spent most of the last fifteen years in the privileged position of a nationally known project, thus being able to observe similar projects succeed and fail for various reasons. Even though we do not have a rigorous analysis of the causes for success and failure in such projects, we observe that failure is more common among projects which have started with a definite tax collection bias, while success rates seem to be higher among multidisciplinary and multipurpose projects.

While empirical, this observation enables us to say, from experience, that much of the driving force behind Belo Horizonte's GIS has originated in the demand posed by real applications, conceived and implemented to address real needs. That was the result of a combination of sensible thinking on the part of the managers that started and supported the project in its early stages, and clearly established paradigms, such as no-cost data distribution, extensive partnerships, and diversity of applications. We observe that this course of action can equally apply to both transitional and developed economies, since its success is mostly due to reaching wide agreements on principles and practices, always guided by real needs and applications.

The future of GIS-BH is clearly related to the development and deployment of services. In the past, the project has evolved from data sharing using proprietary format files to neutral interchange formats, and then on to Internet-based raw data distribution, and to a database-centric interoperable approach. The combination of the various kinds of demands that GIS-BH currently fulfills naturally leads to a service-oriented SDI (Bernard & Craglia, 2005). In the process, the current role of metadata must be expanded and enhanced, and spatial ontologies can be used to facilitate consensus building on themes related to cooperation and on-line data sharing.

Acknowledgments

The authors would like to thank and acknowledge Prodabel's GIS team for the many years of innovative and cooperative work. Clodoveu Davis' work is partially supported by CNPq, a Brazilian agency in charge of fostering research and development. Frederico Fonseca's work was partially supported by the National Science Foundation under NSF ITR grant number 0219025 and by the generous support of Penn State's School of Information Sciences and Technology. The authors also would like to acknowledge the many helpful comments from the editors of this special issue and from the anonymous reviewers.

References

Bernard, L., & Craglia, M. (2005). In SDI: From spatial data infrastructure to service driven infrastructure. Paper presented at the Research Workshop on Cross-Learning between Spatial Data Infrastructures (SDI) and Information Infrastructures (II), Enschede, Holanda.

Bernstein, R.J. (1983). *Beyond Objectivisim and Relativism: Science, Hermeneutics, and Praxis.* Philadelphia: University of Pennsylvania Press.

Borges, K.A.V., Davis, Jr., C.A., & Laender, A.H.F. (2001). OMT-G: an object-oriented data model for geographic applications. *GeoInformatica* 5(3), 221–260.

Braa, J., Monteiro, E., & Sahay, S. (2004). Networks of Action: Sustainable Health Information Systems across Developing Countries. *MIS Quarterly*, 28(3), 337–362.

Davis, Jr., C.A. (1993). In Address base creation using raster-vector integration (Vol. I, pp. 45–54). Paper presented at the Urban and Regional Information Systems Association Annual Conference (URISA'93), Atlanta, Georgia.

Davis, Jr., C.A. (1995). Interchange of geographical information (in Portuguese) (Technical Report No. 01/1995). Belo Horizonte: PRODABEL.

Davis, Jr., C.A. (2002). Geographical information interchange: the experience on standardization and cooperation in Minas Gerais state (in Portuguese). In G.C. Pereira & M.C.F. Rocha (Eds.), Geographic Data: aspects and perspectives (pp. 43–54). Salvador (BA): Rede Baiana de Tecnologias da Informação Espacial (REBATE).

Davis, Jr., C.A., & Borges, K.A.V. (1994). In Object-oriented GIS in practice (Vol. I, pp. 786–795). Paper presented at the Urban and Regional Information Systems Association Annual Conference (URISA'94), Milwaukee, Wisconsin.

Davis, Jr., C.A., & Oliveira, P.A. (2002). Interoperable and distributed GIS for large local governments (in Portuguese). *Informática Pública*, 4(1), 121–141.

Davis, Jr., C.A., & Zuppo, C.A. (1995). In Updating urban geographic databases: methodology and challenges (Vol. 1, pp. 269–276). Paper presented at the Geographic Information Systems/Land Information Systems 1995 Annual Conference, Nashville, Tennessee.

Fonseca, F.T., & Davis Jr., C.A. (1999). Using the Internet to access geographic information: an open GIS interface prototype. In M.E. Goodchild, M.J. Egenhofer, R. Fegeas & C. Kottman (Eds.), *Interoperating Geographic Information Systems*. Kluwer Academic Publishers.

Fonseca, F.T., Egenhofer, M.J., Davis Jr., C.A., & Borges, K.A.V. (2000). Ontologies and knowledge sharing in urban GIS. *Computers, Environment and Urban Systems*, 24(3), 251–272.

Fonseca, F.T., Egenhofer, M.J., Davis Jr., C.A., & Câmara, G. (2002). Semantic granularity in ontology-driven geographic information systems. *Annals of Mathematics and Artificial Intelligence*, 36(1–2), 121–151.

Fonseca, F.T., & Zuppo, C.A. (1994). In School pre-registration and student allocation (Vol. 1, pp. 30–40). Paper presented at the Urban and Regional Information Systems Association Annual Conference (URISA'94), Milwaukee, Wisconsin.

Gadamer, H.-G. (1975). *Truth and Method*. New York: Seabury Press.

Georgiadou, Y., Sahay, S., & Hanseth, O. (2004). Implementation of Spatial Data Infrastructures (SDI) in Transitional Economies (Technical Report). Omaha, NE: College of Information Science & Technology, University of Nebraska at Omaha.

Habermas, J. (1971). *Knowledge and Human Interests*. Boston: Beacon Press.

Meinberg, F.F. (2003). Tools for the analysis of traffic accidents using GIS (in Portuguese). *Informática Pública*, 5(1), 79–99.

Oliveira, M.P.G. (1997). Spatial decision support systems: models for analyzing the densification of economic activities in urban space (in Portuguese). Unpublished Masters Dissertation, Fundação João Pinheiro, Belo Horizonte (MG).

Oliveira, S.M., Sousa, R.P., & Amaral, F.M. (1995). In A review of the limits of urban census sectors and other urban spatial reference units (in Portuguese) (pp. 857–863). Paper presented at the XVII Brazilian Cartographic Conference, Salvador (BA).

Oliveira, S.M., Sousa, R.P., Davis Jr., C.A., & Amaral, F.M. (1996). GIS in the definition of spatial reference units for Belo Horizonte's urban quality of living indicator (in Portuguese). *EspaçoBH,* 1(0), 21–29.

Peixoto, N.M.A. (2000). Methodology for the creation and compatibilization of a legalized parceling base map for the city of Belo Horizonte (in Portuguese). Unpublished Masters Dissertation, João Pinheiro Foundation, Belo Horizonte (MG).

Pessanha, J.E.M., & Carvalho, F.R. (1999). Creation of an information system based on methods standardization for zoonosis control in the city of Belo Horizonte (in Portuguese). *Informática Pública,* 1(1), 63–74.

Pinto, M.V. (1999). School pre-registration: democratization of the access to public education (in Portuguese). *Informática Pública,* 1(2), 139–156.

Reed, A.M. (2000). Rethinking development as knowledge: implications for human development (Technical Report). Rome: United Nations Office for Project Services.

Sagasti, F.R. (2004). *Knowledge and Innovation for Development: The Sisyphus Challenge of the 21st Century.* Cheltenham, UK; Northhampton, MA: Edward Elgar.

Silva, T.E.P.P., & Ottoni, M.V. (1995). In The importance of cartography for the success of GIS deployment in the city of Belo Horizonte (in Portuguese) (pp. 864–873). Paper presented at the XVII Brazilian Cartographic Conference, Salvador (BA).

Sousa, R.P., & Bretas, N.L. (1994). In GIS in the surveillance of infant mortality of poor areas in a two-million-people city (Vol. 2, p. 1). Paper presented at the Urban and Regional Information Systems Association Annual Conference (URISA'94), Milwaukee, Wisconsin.

The World Bank. (1999). *World Development Report—Knowledge for Development.* New York: Oxford University Press.

Zuppo, C.A., Davis, Jr., C.A., & Meirelles, A.C. (1996). In GIS in the transportation and traffic systems of Belo Horizonte (in Portuguese) (pp. 376–387). Paper presented at the GIS Brasil 96, Curitiba (PR).

10

An Exploration of SDI and Volunteered Geographic Information in Africa

Yola Georgiadou, Nama Raj Budhathoki, and Zorica Nedović-Budić

CONTENTS

10.1 Introduction

Spatial data infrastructure as a notion and a set of practices has made inroads in African countries since the first Committee on Development Information (CODI) conference in 1999 (CODI-I 1999). Still, SDI is fraught with problems and arguably slower to implement in Africa than in more economically developed regions (Georgiadou and Homburg 2008; Homburg and Georgiadou 2009). The most recent official report on SDI activities by African states mentions the lack of policies, the reluctance of governments to implement policy guidelines, and the lack of clarity on roles and responsibilities at national, provincial, and local levels as the major impediments to SDI implementation (United Nations Economic Commission for Africa 2009).

In the same report, other challenges include the very limited funding allocated to SDI development, lack of facilities, obsolete equipment, the incompatibility of data formats among institutions and a majority of data sets still in analogue form. Encouraging signs include a few countries with SDI coordinating bodies and working groups, new mapping initiatives, new mapping revision initiatives, an African metadata profile, and installed GNSS stations for the African reference frame (United Nations Economic Commission for Africa 2009).

The United Nations Economic Commission for Africa (UNECA) has played a critical role in helping build consensus around key African development challenges and in articulating common African perspectives and positions. Its mandate to convene senior policy makers—through the annual Conference of Ministers of Finance, Planning and Economic Development—and other development stakeholders is pivotal in ensuring this role. The Committee on Development Information (CODI), a subsidiary body of UNECA, provides policy and technical guidance for the "harnessing information for development" program. The program's objective is to assist African governments in the fields of spatial data infrastructure (SDI) and national information and communication infrastructure (NICI) development. CODI has regularly convened voting delegates from African member states and international observers in biennial conferences to discuss policy issues related to the implementation of SDIs and NICIs by planners and policy makers. These require a vast amount of geographic information to address the aspirations articulated by African and global initiatives, such as the United Nations Millennium Declaration (http://www.un.org/millennium/declaration/ares552e.pdf).

Appeals for better national data infrastructures are made regularly at these conferences. For instance, at CODI-III it was recognized that "there is a need to review ... indicators in the light of the Millennium Development Goals and explicitly address Millennium Development Goals in the National Information and Communication Infrastructure (NICI) development process" (CODI-III 2003, p. 18). At CODI-IV, the importance of spatial data to African initiatives was underlined: "statistical data, Geographic Information Systems (GIS) and ICTs ... could be instrumental in assisting member States achieve the Millennium Development Goals (MDGs), address Poverty Reduction Strategies (PRS) and supporting the New Partnership for Africa's Development (NEPAD)" (CODI-IV 2005, p. 2). However, progress with implementation in Africa is slow (United Nations Economic Commission for Africa 2009).

Recently, UNECA developed a tool designed for visualizing and analyzing progress of member states toward the millennium development goals, many of which are related to basic public services (CODI-V 2007). Data used in this application are procured from the United Nations (UN). The poor are counted, public service levels are inventoried, and the data are aggregated to national levels and submitted to the UN. Yet concerns about the data are increasing. As UN Habitat (2010) puts it, "Current attempts to improve monitoring approaches have been hampered by the lack of reliable information at the local level, resulting in statistics which mask the true picture on the ground."

Volunteered geographic information (VGI), a recent phenomenon fueled by the emergence of Web 2.0 and geo-browsers, may provide a new alternative to data collection at the local level. "Citizens as voluntary sensors" (Goodchild 2007a) can contribute spatial data about their immediate and intimately familiar environment using Google Earth's application program interface (API). However, citizens' submissions in VGI do not need to be limited to the geometric primitives of point, line, or polygon (Budhathoki 2010). Citizens could participate in this new phenomenon as active partners of government in the process of public policy making, or as "citoyens" (van Duivenboden 2005; van Ooijen 2008). The view of the "citizen as citoyen" has been proposed by public administration scholars to emphasize the democratic value of direct citizen participation in government activities.

The rest of the chapter is organized as follows. In Section 10.2, we highlight the trajectory of SDI development in the African continent in the past 10 years from the point of view of UNECA. In Section 10.3, we outline the emergence of the volunteered geo-information phenomenon and explore two cases of citizen sensing in Africa. Finally, we discuss research opportunities that could advance VGI in Africa.

10.2 SDI in Africa

Already at the first CODI conference in 1999, GIS was singled out for its importance "as economic investment and as a tool to assist in alleviating some of the problems being faced by African countries" (CODI-I 1999, p. 3). SDI as a notion made its official début in 2001, in a position paper of UNECA titled "The Future Orientation of Geoinformation Activities in Africa" (United Nations Economic Commission for Africa 2001). By that time, UNECA's "Action Framework to Build Africa's Information and Communication Infrastructure" was already in full swing of implementation but largely oblivious of spatial data and technologies (United Nations Economic Commission for Africa 2003a).

In 2001 the African Information Society Initiative (AISI) was adopted by the conference of African Ministers of Planning and Finance as the guiding framework for development of national information and communication infrastructure (NICI) plans. NICIs were meant to support decision making at all levels and provide information and communication infrastructure for government, business, and society to enlighten the process of development (United Nations Economic Commission for Africa 2003a).

In 2003, UNECA cosponsored the publication of "SDI-Africa: An Implementation Guide" in response to the member states' need for instruction on how to proceed to set up a national spatial data infrastructure (UNECA 2003b). The guide summarized experiences, examples, and documents from African countries and other countries around the world for information managers, data technicians, and technology innovators interested in building information infrastructure in their African country. The guide was envisioned as "guidelines on concrete steps to implement SDIs in Africa, targeted to all those that have a key role to play in promoting, adopting, developing or implementing spatial information infrastructure in their home countries" upon recommendation of (CODI-II 2001) and reinforced by appeals of representatives from member states, academia, professional bodies, and other sectors (United Nations Economic Commission for Africa 2003b). During the compilation of the guide, the practical difficulty of developing a "how-to" step-by-step guide for something that defies computation was resolved by organizing the material as chess moves and options (United Nations Economic Commission for Africa 2003b).

In 2005, it became obvious that a list of options from which to choose was not enough. "It is not enough to … just provide [policy makers] with a list of starting options to choose from" (Bassolé 2005, p. 35). This recognition precipitated a stocktaking of the elements still missing for SDI development:

> There is no descriptive type of implementation steps provided for countries embarking on SDI development in Africa to follow. The reason put forward to explain this is the evolving nature of SDI. While such a cautious attitude is admirable in a research context, it constitutes an issue in a development environment. Policy makers in Africa cannot wait for a consensus to be reached, or for the SDI concept to stabilize, before being guided in operational terms, on how to develop their national spatial data infrastructures. (p. 35)

With progress still elusive in 2005, African commentators began to portray SDI as an AISI implementation tool in an effort to align the faltering SDI implementation with the successful NICI implementation in Africa. For instance, Bassolé (2005) envisaged two possible SDI/NICI integration solutions: (1) a soft solution, where action leaders and plan implementation managers on both sides exchange information regularly on their respective plans and activities and consult among themselves for policy orientation and follow-up on their progress, and (2) a hard solution whereby actions on both sides are coordinated within the framework of a common geo-enabled NICI plan taking account of the SDI vision and mission. Both solutions, however, are fraught with difficulties:

> In an environment where human resources management issues are important, in particular where the Geoinformation community is an interest group with strong influence, the soft solution may be adequate. However, it has the drawback of difficult coordination, and risk of imbalance in implementation due to difficulties linked to funds mobilization for the Geo sector. From a result-based management point of view, the hard option is the most appropriate, even if it is entailed with fear from the Geoinformation community to be dominated by the ICT sector. (p. 39)

A few months after the completion of Bassolé's report, the CODI-IV conference resolved unanimously to include spatial data in the objectives of AISI (CODI-IV 2005), while attempts to enshrine SDI formally in the AISI constitution remained fruitless.

In 2007, UNECA developed the MDG Mapper, a tool designed for visualizing and analyzing progress of member states toward the MDGs, many of which are related to basic public services (CODI-V 2007). All data used in this application were procured from the official MDG database maintained by the United Nations Statistics Division. Yet concerns about the official MDG database are increasing. One concern is the lack of capacity in most countries to provide relevant, reliable, and up-to-date national statistics (Saith 2006). A second concern is the lack of disaggregated monitoring below the national level. Vandemoortele (2008, 2009), one of the chief architects of the MDGs, notes that disparities within the countries have grown to the point of slowing down national progress.

Yet another concern is that some MDG indicators cannot be taken at face value because they use both observation and assumptions in varying proportions. Vandemoortele (2003) points out that statistics on water availability frequently overstate access to water in urban areas because urban residents within 100 meters of a water point are classified as adequately covered; this assumes that the tap is working and the water is of good quality, which is often not true. Schools and hospitals are duly counted in the official statistics provided to the UN, but the assumption that school teachers and health workers are at work is frequently not true (Chaudury et al. 2006). National-level statistics conceal the true picture on the ground (UN Habitat 2010).

10.3 Volunteered Geographic Information

The emergence of geo-browsers and Web 2.0 opens new possibilities for data collection at the local level. Citizens can act as voluntary sensors of geometric primitives (points, lines, and polygons), interact with each other, provide these spatial data to central sites, and ensure that data are collated and made available to others (Goodchild 2007b):

> A new, exciting, and at the same time problematic alternative ... allows any user to create and publish new content, in the form of layers that can be viewed over the Google Earth imagery base. Tens of thousands of sources, many of them citizens with no prior experience in geographic information technologies, have taken advantage of this mechanism in recent months. (pp. 8–9)

Goodchild (2007a, 2007b) gave the name "volunteered geographic information" (VGI) to this new phenomenon.

Geographic information volunteered by citizens redistributes the right to define and judge the value of the spatial data and facilitates a new system of data coproduction (Coleman, Georgiadou, and Labonte 2009). Visualizations by lay people that find their way into the public discourse or decision making may instigate difficult issues of legitimacy of multiple perspectives, voices, and visual expressions (Sheppard and Cizek 2009). Trust issues can be raised with regard to the absence of metadata for geographic information supplied through Google Earth and other commercial actors who control the data, its quality, and its accessibility (Craglia et al. 2008). Relying on commercial sources for data carries a risk of biased representation of world events (Harvey 2007) since the dates of the images and their resolution reflect perceptions of market potential and not of public interest (Perkins and Dodge 2009). Along with these challenges, there are also opportunities.

The first is related to citizens as sensors of spatial data. While the use of spatial knowledge held by ordinary citizens is not new in itself—citizen input has been used in tasks such as cadastral adjudication, update of topographic maps, driving directions, and travel guides—the ease of use of Web 2.0 based tools and the sense of empowerment people feel from using these tools open new possibilities (Budhathoki, Bruce, and Nedović-Budić 2008). In many parts of the world, even basic information on road conditions and locations of medical services is not readily available. In places where such information is available, access to it may be expensive or restricted. In this context, the engagement of millions of ordinary people in the production, sharing, and creative use of spatial data (Miller 2006) looks promising; it also requires a reconceptualization of an SDI user as "produser" (Budhathoki et al. 2008).

The second opportunity requires a reconceptualization of a "citizen" as "citoyen." Public administration researchers distinguish a number of citizen–government relations, including: (1) citizen as a subject of the state meeting legal requirements, (2) citizen as a client engaging in transactions with government, and (3) citizen as a citoyen directly participating in policy processes (van Duivenboden 2005). Information and communication technology (ICT) can potentially shape each of these citizen–government relations. For instance, location-based information and communication technology may provide the state with opportunity to detect deviant behavior of the "citizen as subject," may enable government to cater to the needs of the "citizen as client," and may facilitate the involvement of the "citizen as citoyen" in processes of decision and policy making (van Ooijen 2008).

In the following sections, we illustrate these opportunities with vignettes from our research in VGI in Africa. The first is a desktop analysis that shows an increasing number of citizens acting as sensors of spatial data in OSM (Open Street Map) in Africa (Section 10.3.1). The second vignette draws from ongoing fieldwork in Zanzibar and explores the role of citoyen in VGI (Section 10.3.2).

10.3.1 Citizens as Sensors of Spatial Data

We use Open Street Map VGI (www.openstreetmap.org) to explore and illustrate the potential of VGI for Africa. We chose OSM because it is frequently cited as a successful VGI project by the GIScience community (Goodchild 2007a; Budhathoki, Nedović-Budić, and Bruce 2010). OSM is also interesting because of its decentralized management and self-organizing approach (Haklay and Weber 2008). OSM aims to collect street network geographic data for the whole world and make it freely available. Thus, the project appeals particularly to those societies that lag in creating and maintaining infrastructure necessary for the production and provision of geospatial data.

The OSM project started in 2004 in London. Africa was mapped for the first time in OSM in October 2006. Nearly 1,000 people took part in mapping Africa between October 2006 and March 2009. Collectively, they mapped about 6 million locations during this period. Citizen sensors have contributed location information in OSM for different parts of Africa. The number of people taking part in VGI mapping is increasing in Africa. As the curve in Figure 10.1 shows, only a few people were contributing each month when the notion of the interactive Web mapping in OSM reached Africa. This number has increased over time with more than 100 people contributing location information by March 2009.

The analysis shows a significant growth of the number of contributors during this period in Africa. Interestingly, we find that 45% of those who have mapped Africa have also mapped at least one more continent. Conceivably, they are Africans living outside Africa or foreigners who have traveled to Africa and have acquired knowledge about the African street network. In either case, our data point to the possibility that those living outside are interested in contributing to the African OSM. In fact, those who live outside may be better positioned for certain tasks because of their access to the Internet and digital devices. Thus, VGI can mobilize people living both within and outside Africa for measuring, mapping, and reporting data about space and situations that concern them.

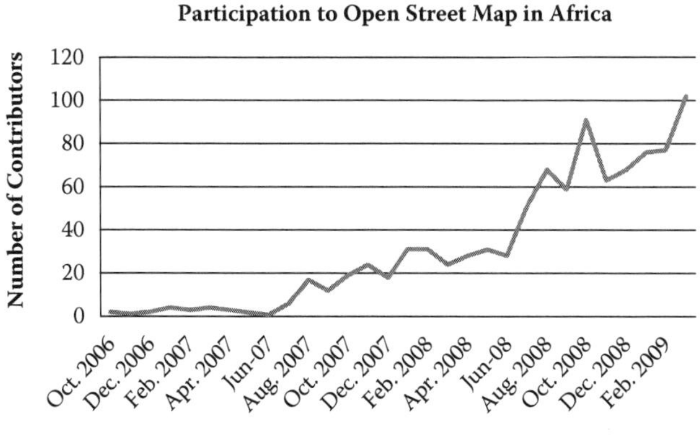

FIGURE 10.1
Increasing trend of citizen participation in mapping Africa.

The emerging VGI infrastructure may help address some of the challenges that African SDIs face. First, nobody owns the underlying technical infrastructure of OSM and hence its use is free. Next, large numbers of technically savvy international volunteers can maintain the infrastructure. Moreover, certain VGI projects such as OSM adopt the "creative commons" licensing approach, allowing anyone to (re)use and (re)mix data freely. Young Kiberans use street networks from OSM for mapping Kibera, Africa's largest slum, in Nairobi, Kenya (http://mapkibera.org/).

The phenomenon of citizens as sensors of spatial data has attracted considerable research attention. In academia, this is evident from special journal issues (*GeoJournal* 2008; *Geomatica* 2010) and a growing number of sessions on this topic in conferences such as the Association of American Geographers and Global Spatial Data Infrastructure. In business, companies are seeking to integrate citizen-contributed geospatial data into their business models. For example, in 2009 Google opened its map for citizens from around the world to edit (Google.com); TeleAtlas sees citizen contribution as a valuable means of keeping its maps current (See "Report map changes" at http://www.teleatlas.com), while CloudMade uses OSM data to develop and commercialize value-added services (CloudMade 2009). However, research is almost exclusively concentrated in the North. Little is known about the adoption, use, and citizen participation in VGI in the South.

10.3.2 Citoyens as Sensors

A citoyen without access to the Internet, but equipped with just a standard mobile phone, can "sense" and report on the conditions of public services such as lack of or contaminated water at public water points and absentee doctors in hospitals and teachers in schools. Citoyens combined in a network together with geo-Web services that publicly disclose these reports on Google Earth can potentially improve government responsiveness to citizens. Communities can apply social pressure on local government and decision and policy makers to improve service delivery. Public authorities are likely to be shamed into action (Pawson 2002) and alert a community that a public service failure has been resolved.

Researchers from the faculty of geo-information science and earth observation (ITC) of Twente University are currently carrying out a pilot "H2.O Monitoring Services to Inform and Empower Initiative—Human Sensor Webs" in Zanzibar, funded by UN Habitat (Nairobi) and Google.org, the philanthropic arm of Google.com (UN Habitat 2009). In Zanzibar, citizens can report the lack or bad quality of water at public water points by sending a simple SMS ("no" for no water and "dirty" for dirty water) to a fixed phone number posted on billboards provided by Zanzibar Telecom (ZANTEL) and placed at 50 public water points (Figure 10.2). The information can be visualized in real time on the Internet, on a Google map (allAfrica.com 2010). The Google map is publicly available and visible by

FIGURE 10.2
Example of the billboards provided by ZANTEL and put at 50 public water points in Zanzibar, advising citizens to SMS the water condition.

other citizens as well as by the water authorities at any Internet cafe in real time (Figure 10.3).

Although it is too early to report definitive results, the idea of "citoyens as sensors" of public services may be promising for Africa and other regions in the South. With the penetration of mobile telephones in Africa (one out of three Africans has access to a standard mobile phone) and wide diffusion of Internet cafes such a system can empower communities to report on service levels and could pressure government to take remedial action and improve public services.

Fieldwork in Zanzibar in September 2009 demonstrated the technical feasibility of the concept (*Daily News* 2010). It also revealed a number of technical and institutional challenges. Users often text long, rambling sentences explaining the water problem, instead of texting a single word "no" or "dirty." Typing errors are common as are spam messages, making some text messages incomprehensible to the client server-based software platform. There is a gender mismatch between the mobile phone owners (usually men) and those who fetch the water from the public water points (usually women).

Community loyalty may override the urgency of reporting the grievance. Community (shehia) members are loyal to their leader (sheha) and

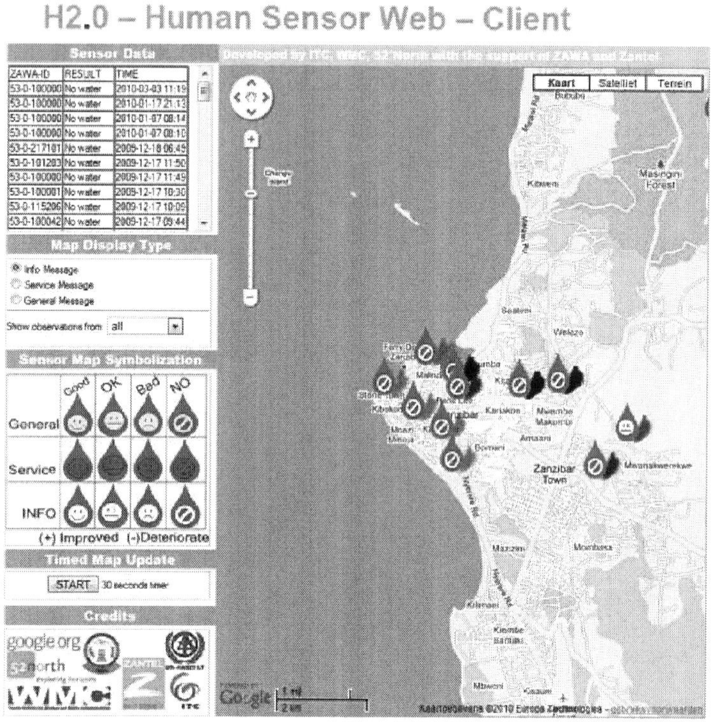

FIGURE 10.3
Real-time, public display of citizens' reports on Google maps.

may refrain from using the human sensor Web unless the community leader endorses it. Distrust of the water authority runs deep in some communities, due to its past history of not responding to citizens' grievances in a timely fashion. Finally, there is the issue of sustainability of the platform. If the platform is scalable—that is, can do more "work" in addition to water (health, education, commercial applications) and can cover larger areas and populations—it is likely to be more sustainable (Walsham, Robey, and Sahay 2007). Thus, issues related to the software platform, to citizens' trust, to government accountability, and to scalability require further research.

10.4 Discussion and Conclusions

We started by discussing SDI development in Africa from the point of view of UNECA, an organization that, more than any other, has promoted the concept of an African Information Society and SDI for Africa. Then we moved on to initial findings of our own VGI research in Africa. The possibilities of contributing to VGI, either directly by citizens as sensors of spatial data or indirectly by citoyens as sensors of the condition of public services, call for increased attention to VGI research in Africa. In one of the chapters of a recent publication of the National Research Council (2010) titled "What Are the Societal Applications of Citizen Mapping and Mapping Citizens?" the authors urge the GIScience community to engage in questions related to:

- The characteristics of producers of VGI and content evaluation
- The ways participation in VGI may have the unintended effect of increasing the digital divide
- Threats to human privacy and new technologies to provide protection

Based on our initial findings, we outline a set of more detailed research questions to explore citizens as sensors of spatial data (citizen mapping) and additional questions on the role of citoyen-sensors in citizen–government relations:

- Citizen-sensors of spatial data
    - A question at the heart of the VGI phenomenon relates to people's motivation to contribute (Budhathoki et al. 2010). Will people continue to contribute information in a sustained fashion or will their initial excitement fade out? How can citizens be motivated to participate in collective information production and sharing? What kinds of incentives would provide for a mix of monetary, hedonic, and social reward for citizen contributors?
    - Other aspects of VGI are also important. How do cost, use restrictions, and availability of digital geographic data through spatial data infrastructures affect the VGI process? Is there a relationship between the diffusion of VGI and the socioeconomic status of a country expressed via the human development index, gross domestic product, literacy rate, and Internet connectivity?
    - VGI has important implications for traditional producers of geographic information. How should national mapping agencies respond to VGI? Which of the framework data layers that the national agencies have been producing should be distributed for citizen production, if any? Which layers should be retained under a centralized supply regime?
    - An equally pressing area of research is the value and potential use of citizen-contributed geospatial data. There is an inherent uncertainty about the quality of VGI data resulting from its openness to external editing. Questions to explore include: Who is using VGI data? How is it used? Why is it used?

- Citoyen-sensors and citizen–government relations
  - Representation methods, frameworks, and models are needed to express formally the physical and application context of the "citoyen as sensor" in the form of a programming language (Broering et al. 2009). How can models for the context of a citizen be formalized in machine-processable language and deployed to understand spatial–temporal patterns in citizen networks of different scale?
  - In order to analyze participation and collective action, we need a better understanding of how different stakeholders distribute their trust within a network of citoyen-sensors (Morawczynski and Miscione 2008). How does citizens' trust in other people, users and designers, technology, organizations, and institutions affect the ways they participate in a network of citoyen-sensors?
  - It is also important to explore how traditional government accountability—reporting up the hierarchy and emphasizing inputs and outputs—can be transformed to responding to citizens and emphasizing results (Dubnick 2005). Under which conditions can a network of citoyen-sensors increase government accountability to citizens?
  - Issues of scalability and sustainability are crucial (Walsham et al. 2007). How does the network mediate collective action in the short and long term? How can it be scaled up to include other public services and serve larger populations and territories? How can its performance be evaluated in terms of citizen empowerment and government responsiveness, as well as in terms of provision of local data to MDGs?

The VGI phenomenon seems to have potential in Africa. Citizens can be more than mere suppliers of spatial data. Citizens with access to standard mobile phones can act as citoyens—direct participants in local government decision making and possibly in addressing aspirations articulated by African and global initiatives, such as the United Nations Millennium Declaration. Research has shown that governments lacking the financial and organizational resources to diagnose social conditions could use people's search practices on the Internet as a proxy. Google Flu Trends (http://www.google.org/flutrends/about/how.html), a not-for-profit project launched by Google.org, illustrates how search engines can anticipate local outbreaks of influenza by counting search engine queries for flu and flu symptoms and *geo-locating* the places where the Internet queries have been made (Ginsberg et al. 2009).

Thus, knowledge claims about societal conditions can be based on people's search practices on the Internet, which search engines capture and analyze *automatically* (Rogers 2009). With VGI, citoyens' *purposeful* action can underpin claims of social conditions (the success or failure in public services) and improve citizen–government relations. VGI research could eventually become part of the African SDI strategy "to foster the development of an indigenous African capability in geospatial science and technology where all the capabilities are maintained and shared by Africans" (United Nations Economic Commission for Africa 2010).

Acknowledgments

We thank Robert Becht and Jeroen Verplanke, leaders of the "H2.O Monitoring Services to Inform and Empower Initiative—Human Sensor Webs" pilot in Zanzibar, as well as Dr. Rob Lemmens and Dr. Gianluca Miscione for the ideas they contributed to this chapter.

References

allAfrica.com. 2010. Tanzania: Want to find safe, clean water in Zanzibar? Just Google it. http://allafrica.com/stories/201002080358.html (accessed August 2, 2010).

Bassolé, A. 2005. Integration of spatial data infrastructures into national information polices: Linking SDI and NICI development processes to speed up the emergence of the African Information Society. Addis Ababa, Ethiopia. UN Economic Commission for Africa, Development Information Services Division.

Broering, A., K. Janowicz, C. Stasch, et al. 2009. Semantic challenges for sensor plug and play. 9th International Symposium on Web & Wireless Geographical Information Systems (W2GIS 2009), December 7–8, 2009, Maynooth, Ireland. *Lecture Notes in Computer Science* (LNCS) 5886: 72–86, ed. J. D. Carswell, S. Fotheringham, and G. McArdle.

Budhathoki, N. R. 2010. Participants' motivations to contribute geographic information in an online community. PhD dissertation, University of Illinois at Urbana-Champaign, Illinois.

Budhathoki, N. R., B. C. Bruce, and Z. Nedović-Budić. 2008. Reconceptualizing the role of the user of spatial data infrastructure. *Geojournal* 72:149–160.

Budhathoki, N. R., Z. Nedović-Budić, and B. Bruce. 2010. An interdisciplinary frame for understanding volunteered geographic information. *Geomatica. The Journal of Geospatial Information, Technology and Practice* 64:11–26.

Chaudury, N., J. Hammer, M. Kremer et al. 2006. Missing in action: Teacher and health worker absence in developing countries. *Journal of Economic Perspectives* 20:91–116.

CloudMade. 2009. http://cloudmade.com/ (accessed August 2, 2010).

CODI-I. 1999. Report of the first meeting of the Committee on Development Information (CODI I): Harnessing information for development. Subcommittee on Geoinformation 28 June 28–July 2, 1999. Addis Ababa, Ethiopia: Economic Commission for Africa.

CODI-II. 2001. Report of the second meeting of the Committee on Development Information (CODI II): Development information and decision making. Subcommittee on Geoinformation, September 4–7, 2001. Addis Ababa, Ethiopia: Economic Commission for Africa.

CODI-III. 2003. Report of the third meeting of the Committee on Development Information (CODI II): Information and governance. Subcommittee on Geoinformation, May 10–13, 2003. Addis Ababa, Ethiopia: Economic Commission for Africa.

CODI-IV. 2005. Report of the fourth meeting of the Committee on Development Information (CODI IV): Information as an economic resource. Subcommittee on Geoinformation, April 23–28, 2005. Addis Ababa, Ethiopia: Economic Commission for Africa.

CODI-V. 2007. Report of the fifth meeting of the Committee on Development Information (CODI V): Employment and the knowledge economy. Subcommittee on Geoinformation, April 29–May 4, 2007. Addis Ababa, Ethiopia: Economic Commission for Africa.

Coleman, D. J., Y. Georgiadou, and J. Labonte. 2009. Volunteered geographic information: The nature and motivation of producers. *International Journal of Spatial Data Infrastructures Research (IJSDIR)* 4:332–358.

Craglia, M., M. F. Goodchild, A. Annoni, et al. 2008. Next generation Digital Earth. *International Journal of Spatial Data Infrastructures Research (IJSDIR)* 3:146–167.

Daily News. 2010. Zanzibar launches water Web-inform project. http://dailynews.co.tz/home/?n=6595&cat=home (accessed August 2, 2010).

Dubnick, M. 2005. Accountability and the promise of performance: In search of the mechanisms. *Public Performance and Management Review* 28:376–417.

GeoJournal. 2008. Special issue on volunteered geographic information. *GeoJournal* 72(3–4).

Geomatica. 2010. Special issue on volunteered geographic information. *Geomatica* 64(1).

Georgiadou, Y., and V. Homburg. 2008. The argumentative structure of spatial data infrastructure initiatives in America and Africa. In *Social dimensions of information and communication technology policy: Proceedings of the 8th International Conference on Human Choice and Computers*, HCC8, IFIP TC 9, Pretoria, South Africa, September 25–26, 2008, ed. C. Avgerou, M. L. Smith, and P. van den Besselaar. IFIP International Federation for Information Processing 282:31–44. Boston: Springer.

Ginsberg, J., M. H. Mohebbi, R. S. Patel et al. 2009. Detecting influenza epidemics using search engine query data. *Nature* 457:1012–1014. http://www.nature.com/nature/journal/v457/n7232/full/nature07634.html (accessed August 2, 2010).

Goodchild, M. F. 2007a. Citizens as voluntary sensors: Spatial data infrastructure in the world of Web 2.0. *International Journal of Spatial Data Infrastructures Research (IJSDIR)* 2:24–32.

______. 2007b. Citizens as sensors: The world of volunteered geography (editorial). *GeoJournal* 69:211–221.

Google Maps. 2009. http://www.google.com (accessed July 29, 2010).

Haklay, M., and P. Weber. 2008. OpenStreetMap—User generated street map. *IEEE Pervasive Computing*. Published by the IEEE CS, DOI 10.1109/MPRV.2008.80:12–18.

Harvey, F. 2007. Just another private–public partnership? Possible constraints on scientific information in virtual map browsers. *Environment and Planning B* 34:761–764.

Homburg, V., and Y. Georgiadou. 2009. A tale of two trajectories: How spatial data infrastructures travel in time and space. *The Information Society* 25:303–314.

Miller, C. C. 2006. A beast in the field: The Google Maps mashup as GIS/2. *Cartographica: The International Journal for Geographic Information and Geovisualization* 41:187–199.

Morawczynski, O., and G. Miscione. 2008. Examining trust in mobile banking transactions: The case of m-PESA in Kenya. In *Social dimensions of information and communication technology policy*, ed. C. Avgerou, M. L. Smith, and P. van den Besselaar. IFIP International Federation for Information Processing 282:287–298. Boston: Springer.

National Research Council. 2010. *Understanding the changing planet: Strategic directions for the geographical sciences.* Washington, DC: National Academies Press.

Pawson, R. 2002. Evidence and policy and naming and shaming. *Policy Studies* 23:211–230.

Perkins, C., and M. Dodge. 2009. Satellite imagery and the spectacle of secret spaces. *Geoforum* 40:546–560.

Rogers, R. 2009. The end of the virtual: Digital methods. Inaugural speech. University of Amsterdam, May 8, 2009. http://www.govcom.org/publications/full_list/oratie_Rogers_2009_preprint.pdf (accessed August 2, 2010).

Saith, A. 2006. From universal values to millennium development goals: Lost in translation. *Development and Change* 37:1167–1199.

Sheppard, S. R. J., and P. Cizek. 2009. The ethics of Google Earth: Crossing thresholds from spatial data to landscape visualization. *Journal of Environmental Management* 90:2102–2117.

United Nations Economic Commission for Africa. 2001. The future orientation of geoinformation activities in Africa: A position paper, ECA/DISD/GEOINFO/DOC/01, Endorsed by the Second Meeting of the Committee on Development Information, Development Information Services Division (DISD). Addis Ababa, Ethiopia.

______. 2003a. An action framework to build Africa's information and communication infrastructure. Geneva: United Nations Office.

______. 2003b. SDI implementation guide for Africa. Endorsed by the Third Meeting of the Committee on Development Information. Addis Ababa, Ethiopia.

______. 2009. CODIST.1—Report on member states' activities since CODI V. Geoinformation Subcommittee, April 28–May 1, 2009, Addis Ababa, Ethiopia.

______. 2010. African regional spatial data infrastructure (ARSDI)—A cooperative geoinformation management in Africa, UN-SPIDER Regional workshop, July 6–9, 2010, Addis Ababa, Ethiopia.

UN Habitat. 2009. H2.O monitoring services to inform and empower initiative—Human sensor Webs. http://www.unhabitat.org/categories.asp?catid=635 (accessed August 2, 2010).

______. 2010. Google and UN Habitat partnership to improve data collection, Zanzibar, January 14, 2010. http://www.unhabitat.org/content.asp?cid=7751&catid=5&typeid=6&subMenuId=0 (accessed August 2, 2010).

Vandemoortele, J. 2003. *The MDGs and pro-poor policies: Can external partners make a difference?* New York: Poverty Group, UN Development program.

______. 2008. Making sense of the MDGs. *Development* 51:220–227.

______. 2009. Taking the MDGs beyond 2015: Hasten slowly. http://www.eadi.org/fileadmin/MDG_2015_Publications/Vandemoortele_PAPER.pdf (accessed August 2, 2010).

van Duivenboden, H. 2005. Citizen participation in public administration: The impact of citizen-oriented public services on government and citizens. In *Practicing e-governmment: A global perspective,* ed. M. Khosrow-Poor, 415–445. Istanbul: Idea Group Publishing.

van Ooijen, C. W. 2008. Territorializing eGovernment: Institutional innovation through the use of location aware technologies. Paper submitted to Study Group 1: Information and Communication Technologies in Public Administration, European Group of Public Administration Conference, September 3–6, 2008, Rotterdam, The Netherlands. http://arno.uvt.nl/show.cgi?fid=95230 (accessed August 2, 2010).

Walsham, G., F. Robey, and S. Sahay. 2007. Foreword: special issue on information systems in developing countries. *MIS Quarterly* 31:317–326.

Section 4

Sense-Making and Summing Up

11

SDI in North and South—A Full Circle Yet?

Gianluca Miscione and Danny Vandenbroucke

CONTENTS

11.1 Introduction

Since the beginning of the 1990s, the notion of spatial data infrastructure (SDI) has inspired different organizations to share geospatial data and achieve seamless integration. After nearly two decades, incidentally overlapping with the Internet age, this vision has gained global acceptance (Homburg and Georgiadou 2009; Masser 2010) and is confirmed by the growing participation in the Global SDI Association's conferences. However, the meaning of SDI in terms of implementation and use has become more fragmented. In spite of a general understanding of SDIs as ways to facilitate access to and use of geospatial data, it is difficult to find wide agreement on a more precise definition of SDI (Grus, Crompvoets, and Bregt 2010).

This becomes obvious from reading the different chapters of this book. All authors agree that SDIs refer not only to the technologies but also to institutional arrangements and practices. Also, all authors argue that the user should be at the center of the SDI and that user needs should be the driving force when designing SDIs. SDIs could underpin local planning practices (Chapters 8 and 9), business processes (Chapter 3), and even citizens as sensors of data and public service delivery (Chapter 10). But how this should be achieved is less obvious from reading the different chapters.

The way we look at SDIs becomes an important issue when a collection of research works includes cases from a global variety of settings from developed and so-called developing contexts. Some readers may question that what we call SDI in India or Uganda is an SDI in Europe. Comparing SDI implementation across contexts assumes that SDI development is a linear process, following preset steps of social and technical change. In fact we can see that within both the Northern Hemisphere (North) and the Southern Hemisphere (South), SDIs take different forms and shapes.

For instance, countries take different approaches in Europe for an SDI initiative like INSPIRE, with its legislation, implementing rules, and guidelines aiming to streamline SDI development, even though the INSPIRE legislation directs them all to evolve toward more active sharing of geospatial data (Vandenbroucke, Janssen, and Van Orshoven 2008). In fact, INSPIRE defines what countries should have (ideally) in common, but leaves in practice a lot of room regarding how this is done and by whom (European Commission 2007).

We are not implying that there are no substantial differences between North and South, but we want to avoid the self-fulfilling prophecy of predefining phenomena in a certain way, only to find what we were looking for and confirm our expectations (Callon and Latour 1981). The relevance is both academic, in the sense of providing a different perspective on SDI as a research domain, and practical, in terms of finding different angles to tackle issues of SDI design and implementation at local, national, and global scales. To rebalance the predominantly North-oriented research, this book compares globally spread cases and draws lessons on the theoretical, methodological, and practical levels.

The open-ended view on how SDIs are used and evolve is probably the most prominent common characteristic of the research studies collected here. These aim to understand and explain SDIs in their actual context before jumping to recommendations about how to improve them. In pursuing an open-ended view, the authors bring social issues into the picture: organizational, interorganizational, economic, legal, cultural, and historical issues. They do so by conceptualizing SDI as enabling resources and distribution mechanisms that cut across organizational levels.

The studies presented here are informed by a variety of methodologies that bridge different disciplines and concepts to understand SDIs as entanglements of social and technical issues. With this enormous challenge in hand, the contributions address the sociotechnical nature of SDI mainly through the lens of social science. A true multi- and interdisciplinary approach to SDI design and implementation and integration of technological and nontechnological disciplines is still to be achieved. However, the coeditors' call for complementing design-prescriptive with theory and empirically grounded explanatory research succeeds in bringing to the foreground usually downplayed issues such as implementation, use, and institutionalization.

We elaborate on these topics by considering four dichotomies: North and South, local and national, social and technical, and explanatory and prescriptive. By discussing where SDI efforts are focused, what SDIs are actually made of, and how different researchers look at them, we discern a complex picture of SDI in real-life settings where its usage and users are not peripheral but rather central in developing successful and sustainable SDIs.

11.2 North and South

The dichotomy of North and South is mainly related to development—a normative concept whose meaning is rooted in the idea of "progress" attached to new technologies and organizational forms. The global acceptance of the northern meaning and manifestations of development has led to transfer of approaches originating in one context (see, for example, the work of Samir Amin on Eurocentrism, 1988) to settings that are incomparable on many accounts. With respect to SDI, we propose to rethink the way it is designed and acted upon in relation to contexts where usual northern assumptions cannot be taken for granted (e.g., developing an SDI based on Internet access or establishing complex coordinating mechanisms). This is well illustrated in Chapter 2 when Silva discusses the institutionalization of the land administration system in Guatemala.

To rebalance the predominantly North-oriented research, this book compares globally spread cases and draws lessons on the theoretical, methodological, and practical levels. Due to their common sensitivity to SDI organizational contexts, the authors show and discuss remarkable differences between the North and the South in terms of what they highlight and what they leave in the background. The analyses from the North assume the availability of geospatial data and focus on the sociotechnical relations along which such data are (expected to be) shared. The studies from the South are more sensitive to geospatial data and their dynamics of production and use.

From an analytical perspective, the North/South dichotomy suggests relevant differences between SDI in the North and in the South. The actual propensity (and possibility) of existing interorganizational relations to be "enacted" (Orlikowski and Scott 2008) by SDI-related efforts needs to be understood. Along this line, Chapters 5 and 8 explain persistence and function of an SDI "antidogma": duplication, redundancy, and proliferation of geo-data and their sources. It must be noted here that although avoidance of duplication efforts is still an important driver for SDI development (as can be seen from the chapters on the developments in the United States and Flanders), redundancy as a motivator is losing ground.*

Even though there are important differences, it is at the same time true that similar problems exist across those regions. In Chapter 4, Nedović-Budić, Pinto, and Warnecke suggest that sharing is still more prominent within organizations than between organizations. Chapter 7 illustrates how, in a well organized project to develop a geo-portal, the original goals were never

* Many SDIs in the North replicate data sets and databases on purpose in order to guarantee continuous access to the data. This is done within the framework of INSPIRE, but it is also applied by such providers as Google. Also, within data sets and databases information might be repeated for practical reasons.

met because dealing with a continuous stream of technological innovations became almost a goal in itself.

Both Northern and Southern contributions concur in challenging the idea that SDIs are drivers of organizational change, and they show how interorganizational relations are a prerequisite rather than the effect of SDI implementation, as convincingly argued in Chapter 2 and as developed in Chapter 3. At this point, we do not want to overlook an important aspect of the dichotomy of North/South: The research on SDI seems to be a northern exercise. In spite of many efforts from the editors, most chapters about the South are not written by southern scholars affiliated with southern organizations. We see this as a relevant research limitation—not because we assume that the locals know more about their own environment, but rather because if their voices are not heard in the academic community, local and tacit knowledge are difficult to access.*

Another limitation resulting from the few southern scholars conducting research on SDI is underutilization of their knowledge in the process of social and political change. A weak presence of local research creates a wider gap between SDI design (taking place in the North), implementation, and use for organizational change (in the South) with often unpredictable results (Rottenburg 2000, 2006). Development projects are considered successful as far as they comply with rules designed and agreed to in the North, with the South having no say in the process. This is much in line with the claims in Chapter 9, which attributes SDI success in Belo Horizonte, Brazil, to the emancipatory character of the project explained by Habermas's typology of knowledge.

11.3 Local and National

SDIs evolve by connecting organizations and people and data and systems into larger artifacts. Rather than starting from scratch, SDIs tend to emerge a posteriori by "connecting the dots." For this reason, local and national levels are not floors upon which to implement SDIs, but rather connection hubs. Different levels contain gateways to link to each other, while each level consists of a multitude of (interconnected) nodes (Vandenbroucke et al. 2009). Connections are across the levels, forming a network that is clearly illustrated with the Flemish case in Chapter 6. Chapter 1 gives a better view, from a legal perspective, on how the local and national levels link to the supranational level—in this case, Europe. It also turns attention to the importance of defining and agreeing on the fundamental concepts that underlie

* Similarly, tacit knowledge enriches SDI research in the North, as evidenced from the work by Harvey and Tulloch (2006) and Nedović-Budić and Pinto (2000, 2001).

multinational SDI activities—primarily, the difficult concept of public task for providing access to spatial data.

As the concept of SDI becomes fragmented because SDIs evolve organically with existing organizational patterns, the long life of SDI vision is noteworthy. The widely accepted myth of SDI (Mosco 2004; Homburg and Georgiadou 2009) mobilizes resources in a variety of contexts, which are gaining relevance in information system research, broadly speaking. Chapter 7 provides an original interpretation of Dutch technological temptation that may be found in Southern settings as well. SDIs span numerous contexts and spread out across multiple scales, even globally—often in precarious balance between global uniformity and local contextual solutions (Georgiadou, Puri, and Sahay 2006). In general, there is a tension between uniformity (and standardization) and specificity, between robustness and flexibility, and between perfect organization (e.g., division of tasks) and ad hoc cooperation.

These issues are also raised in the development of higher level SDIs like INSPIRE (European Commission 2010). Examples are the discussions on what should be the scope of data specifications (host the wishes of everyone in the data model against the choice for a core that suits many) and who should be involved in the process (only key stakeholders or all interested and relevant stakeholders as promoted by the democratic spirit of Janssen, Crompvoets, and Dumortier in Chapter 1).

Universalist views are unlikely to provide exhaustive explanations of how SDIs gain and sustain their dynamic; therefore, context has to come into the picture. The usual way to consider context is to classify it according to levels (Avgerou and Walsham 2000). From the chapters in this book, however, we see how the subnational level becomes more important, as is illustrated by the Flemish case (Chapters 3 and 6). Also across the national level, it is equally important to take local specificities into account (see the case of the United States in Chapter 4). This is supported by other SDI literature (Rajabifard et al. 2006; Masser 2010; Nedović-Budić et al. 2009).

When coordination across organizations, groups, and individuals is not facilitated by colocation, SDIs are expected to keep data-related activities aligned. This is the common rationale for SDIs cutting across levels. This becomes clear in the social network analysis in Chapter 6. While the SDI in Flanders includes mainly actors from the Flemish (subnational) level, the network is clearly interwoven with nodes from the federal (national), provincial, and municipal levels, as well as with other nodes at the European level. SDIs literally do not stop at borders.

To expand this understanding, we note also that the contexts of reference for SDI cannot be sliced according to predefined levels. The settings within which the majority of actors operate and their actions are legitimated are not necessarily hierarchical. Local and national levels are affected by international policies, or just technological trends, in nonlinear ways. For example, a new good practice from a different country can be adopted at the local level without being fostered by the national level. Following Callon and Latour's (1981) advice against creating micro/macro distinctions a priori, we do not assume the explanatory relevance of the local/national level before looking at the actual contexts of reference and legitimization for SDI.

This nonhierarchical view is in consonance with Czarniawska-Joerges and Sevon (1996), who conceptualize the global spread of changes as a continuous transformation of ideas into objects, actions, institutions, and other ideas—ideas spread by continuous processes of embedment and disembedment. From this perspective, the global can be seen as an extended network of localities instead of transcending the local (Czarniawska-Joerges and Sevón 1996). This flat and "embedded in actions" concept of "global" highlights SDI as part of the actual contexts. This is evident in Chapter 10, where citizen-sensors are by default embedded into the variety of contexts—some local and some international—contributing volunteered geographic information long distance.

Davis and Fonseca (Chapter 9) take the local perspective in their case. Chapter 2 (Silva) may be positioned more at the national level. In both studies, the hierarchical logic of a national level deciding and a local level executing does not explain the actual situations, nor does it help deliver useful recommendations. The research in Chapter 8 discusses the logic underlying SDI from the perspective of local planning processes, highlighting a mismatch between local perceptions and activities and SDI implications. Yet another context to zoom in on is the reality on the work floor, where particular organizational structures and business processes should be taken into account in order to make the SDI work (as shown in Chapter 3). Also of interest might be to make a distinction between behavior of organizations and individuals (Wehn de Montalvo 2003; Omran 2007).

11.4 Social and Technical

Paraphrasing Bowker (2000), similarly to other information infrastructures, SDIs operate simultaneously at the concrete level of design and implementation (fields in a database, capacity building, integration of data sets and organizational practices) and at an abstract level (dealing with the relationships between information, organization, service providers, and global software development, among others). "It is vital to dissolve the current disjunction between database (as technical storage medium) and policy (as way of acting in the world). The production of the database is productive of the new world we are creating" (Bowker 2000, p. 676). To a certain extent, this idea is similar to what in the (geo-) standardization world is called the "universe of discourse," which holds a view of the real or hypothetical world including everything of interest (ISO 2002).

Star and Ruhleder (1996) criticize the idea of infrastructure as a "thing." Infrastructures are closely interwoven with people's work and activities, so they exist and evolve in relation to the organized practices that embed them. Therefore, they become invisible unless they break down. Indeed, as our cases show, SDIs emerge differently in relation to who does what, where, and when. The work of one organization is a resource to another. Hence, SDIs are not a thing, but rather a relational entity, as can be seen in Chapters 3 and 6 and is implied in Chapters 4 (the United States), 5 (Uganda), 7 (The Netherlands), and 8 (India) (see also Vandenbroucke et al. 2009).

Star and Ruhleder (1996) suggest asking *when* is infrastructure, rather than *what* is infrastructure. For example, parcel data are mandatory for cadastres and can be used by a finance ministry if technological standards and regulations allow it. The obstacles to this relational functioning of SDI are prominent in the studies about the South, which bring to the front stage the lack of coordinated bureaucratic (Chapter 5), institutional (Chapter 2), organizational, and societal (Chapter 8) relations underneath SDI deployment.

To see this in terms of SDI, making a digital map, for example, requires integration of different layers. Thus, the cartographer and the personnel have to search for the right data, to make several choices about from whom to receive data, the format, and the quality of data. Then, they must (eventually) negotiate access and terms of use with data custodians, make hardware and software platforms available, or use existing (Internet) tools. This variety of actions is dispersed and affected by a high number of actors and technological artifacts that have to be temporarily aligned.

Therefore, in their travel across levels and settings, SDIs are both transformed and transforming in terms of actual functioning, requirements, failures, and responsibilities attached. To grasp the transformational role of SDI in terms of a metaphor, we may say that it is like Kodak, which did not simply invent a new photographic technology, but also created popular photography by mobilizing a huge number of actors and interests.

Looking at SDIs as distributive processes between the technical and the social helps in understanding their dispersed nature as combinations of social, technical, local, and global resources. Relying on a conceptualization of information infrastructures that emphasizes the complexity of large-scale interconnected information systems, Contini and Lanzara (2008) call information infrastructures in complex institutional settings "assemblages." They are heterogeneous sociotechnical networks, not simply technical networks being passively shaped by managers. The physical connections and equipment, technical standards, conventions of use, technical and organizational support structures, organization of work, and cooperation are constitutive parts of the infrastructure.

The growth of assemblages is also shaped by their installed bases, which are constituted by what is already in place at all levels. Thus, the installed base provides both possibilities and constraints for infrastructural evolution. Radical and abrupt changes are rare; intervention attempts need to take into account the inertia or flexibility of the installed base. Not limiting the installed base to its technical dimension helps to understand both the actual constraints described about Uganda (Chapter 5) and Guatemala (Chapter 2) and the actual possibilities of mobilizing millions of citizen-sensors as suggested in Chapter 10. The role of lay people is thus becoming one of great interest: empirically because of the increasing use of geo-ICT in accessing, producing, manipulating, and sharing information and theoretically because of the consequences for a constructionist view, which includes the question of shifts in power between different actors.

11.5 Explanatory and Prescriptive

Descending from the previous discussion about local and national levels and social and technological components of SDI, it is clear that context cannot be reduced to the physical surroundings in which SDIs are situated and where researchers collect data. SDIs need to be contextualized in organizations, rules, technologies, and the skills that actually create, legitimize, and constrain them, which change with time and scale.

In terms of methodology, contributions about the South tend to be based more on qualitative methods, whereas those about the North are more eclectic and rely on both quantitative and qualitative data. An explanation for such a difference can be that measuring phenomena in the South may be more difficult for contingent reasons. Measuring itself—with what it implies in terms of social legitimization and consequent actions—cannot be taken for granted. To provide reliable interpretations, researchers in the South need to be more open to reconceptualizing and changing the assumptions promoted in the Northern sources. For example, in the field of land administration, customary land tenure systems are prominent in many countries and often conflict with geometry and legal enforcement inscribed in cadastral systems (De Soto 2000). This is an important difference with most countries in the North. In general, in the North measurement maintains a paramount role (Bouckaert and Halligan 2008; Van Dooren, Bouckaert, and Halligan 2010).

This is not to say that quantitative measures could not be developed and used for studying the SDI phenomenon in the South. In fact, it is more a question of establishing appropriate measures that reflect the local context and circumstances than of deeming the processes in the South as nonmeasurable. Quantitative and qualitative approaches are both valuable, and their use depends on the nature of the research more than on the location. Chapter 7 on the North and Chapter 5 on the South are the cases in point, respectively, providing the qualitative and quantitative methods suited to the problem and context at hand. The prevalence of qualitative research in the South may be an indicator of the complexities embedded in both problems and contexts that are poorly understood and not easily subjected to available measures. Thus, these rich and largely unexplored dynamics require the depth of analysis and investigative detail that would, over time, possibly lead to the formalized procedures generally used for well defined or more mature phenomena.

Another ambitious goal of this volume is to emphasize the explanatory dimension of research. To ensure understanding of SDI, the actual realm of use has been put in the foreground (infrastructural inversion, Star 1996). It has been pursued by looking at tangible applications in urban planning, decision making, administration, and management. It has also required that each study be theoretically framed and include empirical work that would be used to test the theoretical propositions.

In a doctoral colloquium preceding this book, some of the authors were asked to situate their research on a two-dimensional graph whose axes were explanatory–prescriptive and exogenous–endogenous change (SDI creates organizational change or vice versa). The majority of respondents placed their research in the explanatory–exogenous quadrant, which signifies that they understand SDI as affected by external forces. Thus, connecting the last two dichotomies, the studies collected here are more sensitive to the social dimension of SDI and epistemologically more oriented toward explanation (Orlikowski and Barley 2001). Methodologies have been designed accordingly to fit explanatory research. It has to be noted about "developing contexts" that the Latin American contributions are inclined toward the adoption of Habermas's theory of communicative action as a normative theory on ideal discursive conditions and categorization of knowledge to technical, practical, and emancipatory areas.

If levels of analysis are not predefined and SDI has a networked nature, what are the research empirical boundaries? How does one trace not only the actors, as actor network theory (ANT) recommends, but also the contexts? An option would be to extend the ANT with contextual information, which is chosen in the analysis of Flanders in Chapter 6. Other methodological venues have been explored in the book: infrastructural inversion (Star and Ruhleder 1994), by Richter et al. (Chapter 8); unbounded ethnography (Engeström 2006), by Koerten and Veenswijk (Chapter 7); and the use of the Internet as a tool to know the wider world (Rogers 2009), by Georgiadou, Budhathoki, and Nedović-Budić in Chapter 10. The reason to mention these approaches is that they all promise to enhance research on SDIs that cut across multiple scales and require consideration of both social and technical agency.

11.6 Conclusions

A rigorous definition of SDI would not have allowed the open approach to the field that the reader can find in this book. Nevertheless, we stress that not relying on a strict definition of SDI does not mean that anything can be an SDI. Rather, following Rottenburg (2006), we argue that SDI provides a vocabulary that acts as a metacode (defined as a universal code that appears to be comprehensible in all frames of reference) that allows both practical developments on the ground and research on SDI. Thus, although we do not find universalist conceptualizations of SDI credible, we think that a metacode is good to have for a mixed community of researchers and practitioners.

Most SDI literature asserts that SDIs improve decision making, support good governance, foster social equity and development, support disaster prevention and management, help manage environment and environmental risks, and improve planning and sustainability of local communities as well as large cities. We might temper the expectations of what an SDI should do. We might also need more research in order to confirm such statements by assessing performance and impact (Crompvoets et al. 2008). On the other hand, does this lead to SDI convergence (the central theme of GSDI11)? We do not have a final answer, but we do not see SDI across the globe converging into a seamless, universal SDI. Only theoretically and empirically grounded cases can show how SDIs enable a diverging variety of activities in dispersed settings.

Therefore, given the resilience of the SDI myth (Homburg and Georgiadou 2009) and the considerable variations of its translations into dispersed settings, we could ask where to ground SDI. To answer this question, we cannot refer simply to the geographic locality of SDI initiatives, but should consider their contexts of reference as well. While the contemporary world is getting "infrastructured" also because of SDI, the studies presented here help in gaining understanding beyond SDIs themselves. For example, organizing public administrations and their relations with citizens can be interpreted by tracing and understanding the distributive mechanisms at which SDIs aim.

By presenting and discussing significant cases, we draw some preliminary conclusions about interactions between SDI-related agencies—with their global outreach—and the heterogeneous institutional settings "crossed" by them. Interdisciplinary theory-based analyses (ranging from neo-institutionalism to ANT, from theory of communicative action to hermeneutics) and a mix of methodologies (from statistical and social network analysis to ethnography) have been used to account for the richness of SDIs in their actual dispersed contexts.

It is difficult to give a black or white answer to the question whether we really need different theories and methods for SDIs in the South and in the North. Certainly, we can claim that while SDIs grow to larger scale within and across a variety of organizational settings around the world, SDI research tends to remain focused on North American and Western European environments (with some interesting exceptions that can be seen, for example, in the work of Delgado Fernández and Capote Fernández (2009)). We cannot take for granted that findings originating in these contexts are necessarily relevant everywhere else. Therefore, research crossing those boundaries is a strategically important area of inquiry.

References

Amin, S. 1988. *L'eurocentrisme: Critique d'une idéologie.* Paris: Anthropos.

Avgerou, C., and G. Walsham. 2000. Introduction: IT in developing countries. In *Information technology in context: Studies from the perspective of developing countries,* ed. C. Avgerou, and G. Walsham, 1–8. Aldershot, England: Ashgate Publishing.

Bouckaert, G., and J. Halligan. 2008. *Managing performance: International comparisons.* New York: Routledge.

Bowker G. 2000. Biodiversity datadiversity. *Social Studies of Science* 30:643–683.

Callon, M., and B. Latour. 1981. Unscrewing the big Leviathan: How actors macrostructure reality and how sociologists help them to do so. In *Towards an integration of micro- and macro-sociologies,* ed. K. Knorr-Cetina and A. V. Cicourel, 277–303. London: Routledge & Kegan Paul.

Contini, F., and G. Lanzara. 2008. *ICT and innovation in the public sector: European perspectives in the making of e-government.* New York: Palgrave MacMillan.

Crompvoets, J., A. Rajabifard, B. van Loenen, and T. Delgado Fernández. 2008. Future directions for spatial data infrastructure assessment. In *A multi-view framework to assess spatial data infrastructures,* ed. J. Crompvoets, A. Rajabifard, B. van Loenen, and T. Delgado Fernández, 385–397. Melbourne, Australia: Melbourne University Press.

Czarniawska-Joerges, B., and G. Sevón. 1996. *Translating organizational change.* Berlin: DeGruyter.

Delgado Fernández, T., and J. L. Capote Fernández. 2009. *Semántica especial y descubrimiento de conocimineto para desarollo sostenible.* La Habana: CUJAE.

De Soto, H. 2000. *The mystery of capital: Why capitalism triumphs in the West and fails everywhere else.* New York: Basic Books.

Engeström, Y. 2006. From well-bounded ethnographies to intervening in Mycorrhizae activities. *Organization Studies* 27:1782–1793.

European Commission. 2007. Directive of the European Parliament and the Council establishing an infrastructure for spatial information in the community. Brussels: Commission of the European Communities.

———. 2010. INSPIRE. http://inspire.jrc.ec.europa.eu/ (accessed July 28, 2010).

Georgiadou, Y., S. K. Puri, and S. Sahay. 2006. The rainbow metaphor: Spatial data infrastructure organization and implementation in India. *International Studies of Management and Organization* 35:48–71.

Grus, L., J. Crompvoets, and A. K. Bregt. 2010. Spatial data infrastructures as complex adaptive systems. *International Journal of Geographical Information Science* 24:439–463.

Homburg, V., and Y. Georgiadou. 2009. A tale of two trajectories: How spatial data infrastructures travel in time and space. *The Information Society* 25:303–314.

ISO. 2002. International Standard ISO 19101, Geographic information—Reference model. Geneva: ISO.

Masser, I. 2010. *Building European spatial data infrastructures.* Redlands, CA: ESRI Press.

Mosco, V. 2004. *The digital sublime: Myth, power and cyberspace.* Cambridge, MA: MIT Press.

Nedović-Budić, Z., G-J. Knaap, N. R. Budhathoki, and B. Cavrić. 2009. NSDI building blocks: Regional GIS in the U.S. *Journal of the Urban and Regional Information Systems Association* 21:5–23.

Omran, E. El-S. 2007. Spatial data sharing: From theory to practice. PhD dissertation. Wageningen, The Netherlands: Wageningen University.

Orlikowski, W., and S. R. Barley. 2001. Technology and institutions: What can research on information technology and research on organizations learn from each other? *MIS Quarterly* 25:145–165.

Orlikowski, W., and S. Scott. 2008. The entangling of technology and work in organizations. London School of Economics and Political Sciences, Innovation Group, Working Papers Series 168.

Rajabifard, A., A. Binns, I. Masser, and I. P. Williamson. 2006. The role of sub-national government and the private sector in future SDIs. *International Journal of Geographical Information Science* 20:727–741.

Rogers, R. 2009. *The end of the virtual–digital methods.* Amsterdam: Amsterdam University Press.

Rottenburg, R. 2000. Accountability for development aid. In *Facts and figures. Economic representations and practices, Jahrbuch Ökonomie und Gesellschaft 16,* ed. H. Kalthoff, R. Jürgen Wagener, and H. Jürgen Wagener, 143–173. Marburg, Germany: Metropolis.

———. 2006. Code-switching, or why a metacode is good to have. In *Global ideas: How ideas, objects and practices travel in the global economy,* eds. B. Czarniawska and G. Sevon, 259–274. Copenhagen: Copenhagen Business Press.

Star, S. L., and K. Ruhleder. 1994. Steps towards an ecology of infrastructure: Complex problems in design and access for large-scale collaborative systems. In *CSCW 94 ACM Conference on Computer Supported Cooperative Work,* 253–264. New York: ACM Press.

Star, S. L., and Ruhleder, K. 1996. Steps Toward an Ecology of Infrastructure: Design and Access for Large Information Spaces. Information Systems Research, 7, 111–133.

Vandenbroucke, D., J. Crompvoets, G. Vancauwenberghe, E. Dessers, and J. Van Orshoven. 2009. A network perspective on spatial data infrastructures: Application to the sub-national SDI of Flanders. *Transactions in GIS* 13:105–122.

Vandenbroucke, D., K. Janssen, and J. Van Orshoven. 2008. INSPIRE state of play. Development of the NSDI in 32 European countries between 2002 and 2007. In *Proceedings of GSDI-10 Conference, Small Island Perspectives on Global Challenges: The Role of Spatial Data in Supporting a Sustainable Future.* St. Augustine, Trinidad, 22 pp.

Van Dooren, W., G. Bouckaert, and J. Halligan. 2010. *Performance management in the public sector.* London: Routledge.

Wehn de Montalvo, U. 2003. In search of rigorous models for policy-oriented research: A behavioral approach to spatial data sharing. *Journal of the Urban and Regional Information Systems Association* 15:19–28.

Conclusion: Implications for Future Research and Practice—Toward Scientific Pragmatism

Zorica Nedović-Budić, Joep Crompvoets, and Yola Georgiadou

The path toward successful spatial data infrastructures, keeping pace with societal and technological developments and issues, is still long, and obstacles are apparent in both the developed and developing world. The appearance of a diffuse set of difficulties is increasingly acknowledged by practitioners in the field, and barriers for implementation have been regularly studied and commented on by academics working from various disciplinary perspectives. While the awareness of barriers regarding development and functioning of SDIs is increasing, this is not to say that they are generally well understood or defined, let alone addressed in both research and practice. This book's objective is to advance the scientific discourse and contribute to SDI practice. More specifically, it attempts to redress the following limitations:

- North-centric nature of most academic and professional accounts in literature
- Emphasis on the national-level SDIs and neglect of the local context
- Technical focus and weak theoretical grounding of scholarly endeavors
- Lack of methodological diversity and rigor

The contributors to this volume, most of whom are PhD students or recent graduates, were recruited to provide work that is interdisciplinary and theory based, but also complemented by methodologically sound empirical inquiry. The contributions were intended to reflect research in the North and South and to employ a range of research methods—quantitative and qualitative—and epistemologies—positivistic and interpretive. Finally, the studies were to reflect various contexts of SDI implementation: local, regional, national, and multinational. In the course of the book's preparation, we discovered that high-quality output that would satisfy the requirements was difficult to find and therefore resorted to three reprints: two for improved regional coverage of the South, which is particularly devoid of research attention, and one for representation of the local level in North America.

We realize that our aspirations were ambitious and could not be easily achieved, but we hope that we managed to provide an opening into important and useful areas of inquiry and ensure visibility of exemplary research efforts. The Summary Table highlights the contents, approaches, findings, and implications for policy and practice of the 10 research chapters balanced between South and North. Brazil, Guatemala, India, Uganda, and Zanzibar (Africa) represented the South; Belgium, the Netherlands, the United States, and the European Union represented the North. These included local levels of analysis, like the cities of Mugdali, in India, and Belo Horizonte, in Brazil; they drew on a multiplicity of theoretical perspectives (e.g., information infrastructures, organizational science, planning theory, sociology, actor-network theory [ANT], and public administration, among others). Finally, they also employed a variety of methods and analytical approaches, including survey, case study, ethnography (narrative), social network analysis, and interpretive method, to mention a few.

The research aim of the majority of the studies is explanatory, although there are a few exercises of exploratory character. This particularly applies to the chapter on exploring the role of citizens in volunteered geographic information (VGI), where the novelty of the phenomenon and related lack of substantial empirical work to date called for a chapter concentrated on a discussion of the trends and prospects in the context of Africa with two vignettes. In this case, introducing the potential of a new technological setup (combination of Web 2.0, geo-browsers, and mobile phone data communication functions) to enhance the provision of local services and empower the citizens to interact more directly and effectively with public agencies took precedence over the theory-testing requirement. Indeed, this more descriptive approach characterizes many early sources on SDIs, with an important role of alerting to trends and setting future research agendas.

The translation of academic research findings to useful pointers toward effective SDI practice is another important objective that has been attempted in this book. The findings of the research presented in this volume carry important implications for SDI policy and practice, as suggested in the Summary Table. Probably the key points have to do with the importance of understanding the local settings; in the case of the South (developing countries in particular), these tend to be more complex and dynamic and include a complicated entanglement of formal and informal practices that do not easily fit the northern models and frameworks for research or practice.

While the divisions and differences do not warrant exclusively separate approaches, flexibility, openness to nonstandard solutions, transparency, bottom-up implementation, application focus, respect for local practices, maintenance of stability and certainty, and empowerment are probably the key underlying principles of a context-sensitive practice. Clearly, these points could be further elaborated in a practice-oriented manual. Translation of research findings to practice is a difficult, time-consuming, and often overlooked task that we would strongly advocate for, particularly for research

SUMMARY TABLE

Contents, Approaches, and Findings of the 10 Book Chapters

Chapter topic/region/ level (Area)	Theory/Field (Key Concepts)	Method/Approach	Findings/Implications for Practice
1. Public task of spatial data provision North Multinational (Europe)	Law, governing, market failure, public good, public interest	Qualitative (conceptual) Logic, argumentation, literature-based overview	Proposes reducing the uncertainty of public task by focusing on specific (spatial information) services and involving stakeholders in a democratic consensus process; Suggests attempting agreement at international and global levels (recognizing political nature of public task/interest); Alerts to the need to balance (spatial) information society and information market
2. Institutionalization of land administration system South National (Guatemala)	Sociology of translation, actor-network theory (ANT), interorganizational cooperation, critical theory	Qualitative (interviews, observation) Interpretive analysis ("telling-showing-telling")	Develops a model for approaching institutionalization (process of "translation"); Identifies key implementation tasks: facilitate cooperation; provide benefits; understand cultures, interests, and power relations
3. Integrating spatial information and business processes: the role of organizational structures North Provincial (Limburg and West Flanders, Belgium)	Organizational (structure, process, information flows, functional concentration)	Mixed quantitative/ qualitative (survey, interviews) social systems approach	Confirms relationship between organizational structure and spatial information use and policy; Suggests that less bureaucratic and centralized divisions of labor could contribute to higher performing integration of spatial information flows in the business processes; Finds that lower level of functional concentration coincides with higher scores for spatial information use and policy

(Continued)

SUMMARY TABLE (Continued)

Contents, Approaches, and Findings of the 10 Book Chapters

Chapter topic/region/ level (Area)	Theory/Field (Key Concepts)	Method/Approach	Findings/Implications for Practice
4. GIS database development and exchange: interaction mechanisms and motivations North Local (cities and counties, United States)	Interorganizational relationships, coordination, interaction (reasons, structures, mechanisms)	Quantitative (survey) Descriptive statistics, difference of means (t-test), cross-tabulation/chi square statistics	Finds that external relationships are more driven by common goals and missions, but also tend to focus on financial resources; internal are also motivated by saving resources (more in terms of time and staff), but also respond to directives and emergency management issues; Asserts data as the main currency of exchanges—external more likely to involve fees, to be legally formalized and facilitated by common standards and clearinghouses; Confirms that informal relationships are important across all settings
5. SDI reality in Uganda: coordinating between redundancy and efficiency South Local to national (cities of Kampala and Entebbe, Uganda)	Resource dependency theory (power and uncertainty) Institutional–public sector inertia	Mixed—quantitative/ qualitative (workshop/ inventory of data and agencies, survey, interviews, focus groups) Axial coding technique relative to context, causal and intervening conditions, action strategies and consequences	Suggests that power loss and uncertainty are the main concerns for organizations (hence, they prioritize single mandates over cross-organizational efficiencies); Finds that large number of donor-funded projects contribute to both redundancy and heterogeneity in data quality (consequently, new redundancies because data users avoid dependency on single data providers as long as the quality is uncertain); Requires matching between SDI coordination objectives and context
6. Social network analysis of the SDI North Local to regional (municipalities, provinces, and the region of Flanders, Belgium)	Public administration Organizational (networks and flows)	Quantitative (survey) Social network analysis (density and centralization)	Demonstrates complexity of organizational interactions—mix of hierarchical and network arrangements; Promotes discovery of complementarities and overlaps (inefficiencies, redundancies)

7. Temptation of technology in NSDI implementation North National (The Netherlands)	Infrastructure implementation (environment, space, and storyboard; stability vs. change)	Qualitative (ethnography, interviews, participant observation) Narrative—interpretation of meanings and sense-making	Distinguishes between requirements of infrastructure and innovation; Suggests avoidance of frequent (technology-driven) redefinition of project goals, assessment rules, and results ("moving target" syndrome)
8. Enlisting SDI for urban planning: local slum declaration practices South Local (city of Mugdali, India)	Public administration/planning, infrastructural inversion, knowledge practices	Qualitative (artifacts—documents and maps, semistructured interviews and informal conversations, field observations) Exploratory Interpretive	Concludes that the role of SDI as an ordering mechanism needs to be evaluated against local practices; Proposes that a more flexible and open approach to SDI may be needed if local practices are to be incorporated; Suggests that the focus on ICT capacity should be complemented with considerations of local processes and stakeholders; Recommends the use of "infrastructuring" approach
9. Considerations from the implementation of a local spatial data infrastructure South Local (city of Belo Horizonte, Brazil)	Critical theory (Habermas's emancipatory knowledge), Gadamer's hermeneutic approach to knowledge (techne and pronesis—borrowed from Aristotle)	Qualitative (interviews, observations) Chronology, description, explanation building	Finds the bottom-up, application-driven implementation as key to success; Requires involvement of multidisciplinary design and implementation team; Affirms funding, cooperation, and political visibility as important implementation factors
10. An exploration of SDI and volunteered geographic information in Africa South Continental to local (Africa, Zanzibar)	Public administration (citizens–government relations)	Mixed quantitative/qualitative Literature-based overview, exploratory descriptive statistics, research agenda setting	Explores alternative approaches to SDI using readily available personalized technologies (e.g., mobile phones); Promises for engagement of citizens in a variety of roles: data development (contribution of geometric primitives), service reporting, and input into local decision and policy processes

that aspires to make an impact. Hence, we call for "scientific pragmatism" that takes societal circumstances and practical needs as the main input in formulating research agendas and considering theoretical (conceptual) frameworks and approaches.

The implications for future research are not radical because many of the gaps already identified several years ago by Budhathoki and Nedović-Budić (2007) have not been fully addressed yet. The authors point to the following areas of SDI that need substantial attention: *definition and conceptualization, inclusive organizational models, standards, monitoring and evaluation, balancing the technical and the social, politics and policy, and a multi-/interdisciplinary approach.* We would like to single out the difficult task of integrating social and technical approaches and issues. This task is pursued in several chapters on the South and the North, although still within the boundaries of social science.

Sociotechnical approaches are particularly relevant in developing countries because of the significant difference in social–political–historical–institutional conditions between the site where the technology is usually designed (a developed country) and the developing country site where the implementation is supposed to take place. Also, the internal and external organizational environments tend to be more complex, sometimes with idiosyncratic intersections of formal institutions and structures with local informal systems, practices, and customs. In many cases, the nature of the problems is different and often not encountered by scholars and practitioners in the North. The power of sociotechnical approaches has been emphatically established, both theoretically and empirically, in the study of implementation of information systems in organizations (Walsham 1993), as well as in understanding the implementation dynamics of information infrastructures that span numerous contexts spread out globally (Rolland and Monteiro 2002).

To elaborate on the importance of contextual and comparative research, Georgiadou, Puri, and Sahay (2005) identify three elements to comprise the basis for a more inclusive and comprehensive research agenda:

- increasing focus on the dynamics of implementing SDIs
- examining the nature and also process of developing and applying standards
- expanding the scope of the design process

The *dynamics of implementation* require opening up the "black boxes" of SDIs and analyzing their interconnections with different constituents of the network and within a historical context that emphasizes both the challenges and opportunities posed by the lock-in effects of the existing installed base. In addition, moving beyond grand visions of what SDIs are, or should do, in normative and prescriptive terms, there is a need to focus on their implementation process, keeping in mind the additional difficulties in institutionalizing SDIs. In SDIs, the complexity is magnified in comparison with traditional information systems/information infrastructures, and the need is for sustained research, well grounded in theory, and employment of a variety of methods, including longitudinal studies.

With stark contextual differences discovered at all levels, the *process of standardization* also should be more sensitive and reflexive, moving from the top-down, "one size fits all" approach and the quest for universal standards that is well entrenched in the information infrastructure research. The authors also suggest that the *scope of the design process* needs to be expanded in terms of who participates and the focus of the effort. A cultivation approach to design, as argued for information infrastructure researchers, needs to start by first identifying the nature of the existing installed base in both technical (e.g., existing maps and their scales) and institutional (e.g., who owns them and their distribution policies) terms. Respecting the functional elements of the existing systems and practices, identifying connections and gateways (nodes) between different parts and actors in the network, and encouraging participatory approaches in the design process also are urged by the authors.

Transpiring from the majority of the contributions is the importance of maintaining a strong user perspective (Masser, Rajabifard, and Williamson 2007) and evaluating SDIs from the user perspective (Nedović-Budić, Pinto, and Budhathoki 2008). While a variety of frameworks is available (Crompvoets et al. 2008; Lance, Georgiadou, and Bregt 2009) between the concepts of usability and utility, the latter certainly offers a relevant standpoint for studying large-scale infrastructures such as SDIs. The user perspective, in general, has reached its time. Gurstein's (2003) framework of effective use of information resources is applicable to SDIs. It reveals that other important organizational and social structures underlie SDIs that enable or limit them. The lens of effective use thus allows us to see SDIs beyond the current paradigm of provision and access of geospatial information. Embedding, mutual adaptation, adjustment, and reinvention of both the technical and the social realms are essential processes that lead toward effective use of information infrastructure. In Stewart and Williams's (2005) words:

> Design outcomes/supplier offerings are inevitably unfinished in relation to complex, heterogeneous, and evolving user requirements. Further innovation takes place as artifacts are implemented and used. To be used and useful, ICT artifacts must be "domesticated" and become embedded in broader systems of culture and information practices. In this process artifacts are often reinvented and further elaborated. (p. 2)

We suggest attention to SDI use and users as another key ingredient of future research agendas across the South and the North. The SDIs promise to enhance governance and decisions, sustainability (environmental, economic, and social, and overall quality of life) can only be achieved through its integration with a variety of user communities and applications. In addition, the comparative perspective is necessary if mutual learning and exchange of experiences are the goals in the emerging international global arena of both science and practice. In fact, the South may provide the ultimate test of the robustness of various theoretical assumptions and frameworks as well as the breeding ground for alternative views and paradigmatic developments. While in this volume the comparative approach is only implicit, the range and variety of settings included provide important insights to justify and facilitate further cross-cultural research and inform international SDI practice.

Many challenges and the urgency for solutions and action (rather than theorizing and research) notwithstanding, we would like to see the wondering nature of SDI and the ambiguity of the SDI notion as an opportunity and a strength rather than a limitation. SDI could join other magic concepts in public management to "play a central role in the … popularization and dissemination of broad ideas of reform. This process affects both academia and the world of practice. It helps to set agendas for both groups. It provides a vocabulary for debate" (Pollitt and Hupe 2009, p. 23). The vagueness of SDIs may also stimulate a new synergy between academics and practitioners, who

> need to acknowledge that these concepts are not precise or even stable—indeed it is in their nature that they are hard to be pinned down and standardized. Magic concepts do tend to attract grants and contracts. This being so, however, they can only fulfill explanatory functions if positioned, specified, operationalized, and applied in systematic ways. Furthermore, less fashionable conceptual equivalents may well be available. For empirical research these may sometimes be preferred to the volatile basis magic concepts seem to provide. (Pollitt and Hupe 2009, p. 23)

In conclusion, while many research gaps related to SDI implementation and institutionalization in the South and the North are still in need of coherent interdisciplinary research efforts, we trust that this book has made one small step in reducing some of the gaps. We would be especially pleased to have provided the communities of SDI scholars and professionals with a sampler of theoretically and methodologically grounded work that also carries pragmatic relevance for SDI implementation and practice in a variety of settings. We also hope that the volume has extended the reach of geographic information science toward other disciplines and fields and has contributed to an enhanced scientific discourse and successful and effective information infrastructures around the globe.

References

Budhathoki, N. R., and Z. Nedović-Budić. 2007. Expanding the spatial data infrastructure knowledge base. In *Research and theory in advancing spatial data infrastructure concepts,* ed. H. Onsrud, 7–31. Redlands, CA: ESRI Press.

Crompvoets, J., A. Rajabifard, B. van Loenen, and T. Delgado Fernandez, eds. 2008. *A multi-view framework to assess spatial data infrastructures.* Wageningen, The Netherlands: Space for Geo-Information (RGI), Wageningen University; Melbourne, Australia: University and Center for SDIs and Land Administration, Department of Geomatics, the University of Melbourne.

Georgiadou, Y., S. K. Puri, and S. Sahay. 2005. Towards a potential research agenda to guide the implementation of spatial data infrastructures: A case study from India. *International Journal of Geographical Information Science* 19:1113–1130.

Gurstein, M. 2003. Effective use: A community informatics strategy beyond the digital divide. *First Monday* 8 (12).

Lance, K. T., Y. Georgiadou, and A. Bregt. 2009. Cross-agency coordination in the shadow of hierarchy: "Joining up" government geospatial information systems. *International Journal of Geographical Information Science,* 23:249–269.

Masser, I., A. Rajabifard, and I. P. Williamson. 2007. Spatially enabling governments through SDI implementation. *International Journal of Geographical Information Science* 21:1–16.

Nedović-Budić, Z., J. K. Pinto, and N. R. Budhathoki. 2008. SDI effectiveness from the user perspective. In *A multi-view framework to assess spatial data infrastructures,* ed. J. Crompvoets, A. Rajabifard, B. van Loenen, and T. Delgado Fernandez, 273–303. Wageningen, The Netherlands: Space for Geo-Information (RGI), Wageningen University; Melbourne, Australia: University and Centre for SDIs and Land Administration, Department of Geomatics, the University of Melbourne.

Pollitt, C., and P. Hupe. 2009. Talking governance: The role of magic concepts. Paper presented at the 2009 conference of the European Group for Public Administration, Saint Julian's, Malta, September 2–5, 2009.

Rolland, K., and E. Monteiro. 2002. Balancing the local and the global in infrastructural information systems. *Information Society* 18:87–100.

Stewart, J., and R. Williams. 2005. The wrong trousers? Beyond the design fallacy: Social learning and the user. In *User involvement in innovation processes: Strategies and limitations from a socio-technical perspective,* ed. H. Rohracher, 9–35. Munich: Profil-Verlag.

Walsham, G. 1993. *Interpreting information systems in Organizations,* Chichester: Wiley.

Afterword

Spatial data infrastructures (SDIs) as typically defined throughout the chapters of this book refer to network-based solutions to provide easy, consistent, and effective access to geographic information and services by public agencies and others. Technical definitions sometimes focus on which minimal capabilities, standards, and offerings of data and services a networked or centralized geospatial processing environment must offer to rise to the level of constituting an SDI. Spatial data infrastructures might also be viewed as virtual spaces built to reflect real space in order to improve decision making in the real worlds in which we live and interact.

Whether it is a networked solution, information-processing environment, or virtual space, the term "infrastructure" in these definitions often infers a foundational role for government. This foundational role for government will remain extremely important even though and while the roles of government agencies, commercial firms, professional and nongovernment organizations, and individuals are changing and evolving in the development of SDIs to support individual and organizational local-to-global decision making.

The chapters contained in this book hold to this assumption. That is, they assume typically that governments will continue to have important and vital roles in the future in facilitating access to core spatial data and services in meeting their needs and those of their citizens, regardless of changing technological and socioeconomic conditions. This is probably a very good assumption. Evolving information marketplace and volunteerism trends are unlikely to replace the public goods components of meeting government needs and ensuring broad-based access to geospatial data and services. Government investment and leadership in meeting modern public and private needs for geospatial information will remain critical.

Governments and their citizens generally benefit by governments continuing to pursue, advance, and adapt development of SDIs. However, the socioeconomic and technological landscape varies from jurisdiction to jurisdiction and that landscape is continually shifting such that governments need regularly to reevaluate the SDI capabilities they should supply and those that might be better supplied by other sectors. Reasonable solutions in a European context are very different from those in a North or South American context and are even more different in African and Asian contexts.

A reasonable balance in societal provisioning of SDI components in one jurisdiction may differ radically from another and the balances for both may be different in a few years due to technological, socioeconomic, and political shifts. Having said this, many system development and sustainability principles are generalizable across jurisdictions, as illustrated through examples and studies contained in this volume. Further, many technological concepts supporting SDIs as developed over the past couple of decades are relatively stable.

In an earlier book involving several of the same authors and editors, a multiview framework for assessing SDIs was presented and explored (Crompvoets et al. 2008). That contribution addressed topics ranging from theoretical foundations to practical assessment methods. This latest book is an excellent follow-up in that it pulls together explicit examples, experiences, and case studies drawn from a range of national and societal contexts. A few of the chapters are valued articles drawn from previously published literature; other chapters are new offerings providing new insights. Practical ways forward in implementing and sustaining SDIs, based on both theory and experience, are provided in several of the chapters.

For those seeking understanding of the factors, processes, and approaches in implementing and sustaining SDIs in complex institutional, organizational, and differing societal settings, the text provides a solid set of readings with which to begin that exploration. Thus, the volume provides a welcome addition to the researcher's bookshelf.

Professor Harlan Onsrud, University of Maine
Executive Director and past President of the Global Spatial Data Infrastructure (GSDI) Association

Reference

Crompvoets, J., A. Rajabifard, B. van Loenen, and T. Delgado Fernandez, eds. 2008. *A multi-view framework to assess spatial data infrastructures.* Wageningen, The Netherlands: Space for Geo-Information (RGI), Wageningen University. Melbourne, Australia: University and Center for SDIs and Land Administration, Department of Geomatics, the University of Melbourne.

Index

U

V

Z